Studies in Systems, Decision and Control

Volume 386

The series "Studies in Systems, Decision and Control" (SSDC) covers both new developments and advances, as well as the state of the art, in the various areas of broadly perceived systems, decision making and control–quickly, up to date and with a high quality. The intent is to cover the theory, applications, and perspectives on the state of the art and future developments relevant to systems, decision making, control, complex processes and related areas, as embedded in the fields of engineering, computer science, physics, economics, social and life sciences, as well as the paradigms and methodologies behind them. The series contains monographs, textbooks, lecture notes and edited volumes in systems, decision making and control spanning the areas of Cyber-Physical Systems, Autonomous Systems, Sensor Networks, Control Systems, Energy Systems, Automotive Systems, Biological Systems, Vehicular Networking and Connected Vehicles, Aerospace Systems, Automation, Manufacturing, Smart Grids, Nonlinear Systems, Power Systems, Robotics, Social Systems, Economic Systems and other. Of particular value to both the contributors and the readership are the short publication timeframe and the world-wide distribution and exposure which enable both a wide and rapid dissemination of research output.

Indexed by SCOPUS, DBLP, WTI Frankfurt eG, zbMATH, SCImago.

All books published in the series are submitted for consideration in Web of Science.

More information about this series at http://www.springer.com/series/13304

Dipankar Deb · Jason Burkholder ·
Gang Tao

Adaptive Compensation of Nonlinear Actuators for Flight Control Applications

Springer

Dipankar Deb
Institute of Infrastructure,
Technology, Research and Management
Ahmedabad, Gujarat, India

Jason Burkholder
Barron Associates
Charlottesville, VA, USA

Gang Tao
Electrical Engineering
University of Virginia
Charlottesville, VA, USA

ISSN 2198-4182 ISSN 2198-4190 (electronic)
Studies in Systems, Decision and Control
ISBN 978-981-16-4160-2 ISBN 978-981-16-4161-9 (eBook)
https://doi.org/10.1007/978-981-16-4161-9

This Springer imprint is published by the registered company Springer Nature Singapore Pte Ltd.
The registered company address is: 152 Beach Road, #21-01/04 Gateway East, Singapore 189721, Singapore

Preface

Actuation is an indispensable aspect of flow control. New actuation devices can potentially lead to the rapid deployment of advanced control system applications. However, there is a need for appropriate compensation of nonlinearities inherent in such actuators. One such type of actuating device is the synthetic jet actuators which have seen active interest in the last couple of decades for next-generation flight control. In this context, many leading research organizations like the NASA Langley Research Center have invented several serial high-performance, piezo-electric, and hybrid synthetic jet actuators to provide active flow control for wing-borne vehicles in subsonic, transonic, and supersonic flow. Efficient flow control results in reduced vehicle drag and the prevention of events that can lead to catastrophic loss. Understanding the correlation between the actuator attributes and resulting flow field, ascertaining the effect of interactions of synthetic jet actuators with the turbulent flow, and determining the correlation of the actuator attributes with surface pressure distribution for control law development are all important aspects in this work.

The traditional objectives of an aircraft design mandate an adequate aerodynamic performance for the entire flight envelope. However, for military aircraft, with unconventional design requirements, the desirables are much more diverse. With the advancement in stealth aircraft, the need to have a low level of radar observability creates an environment where the importance of vehicle geometry and the need to optimize it precedes aerodynamic ruggedness.

Synthetic jet actuators produce a jet flow using the surrounding air rather than relying on a secondary fluid. Synthetic jets exhibit a highly nonlinear relationship between their different parameters. Synthetic jets are currently being investigated for applications such as virtual shaping of jet engine intakes and subsonic projectiles. The core of this approach is establishing a feedback loop between new computational models, lab tests, and field experiments in order to mature the actuation system design in a time-efficient and cost-effective and ready-implementable manner. A model-based environment is needed for the advancement of design and performance validation of synthetic jet actuators. However, formulating a comprehensive, wind tunnel-validated synthetic jet model is technically challenging and expensive.

Given the varied and growing list of synthetic jet applications, adaptive control can cancel the actuator nonlinearities to achieve performance objectives without requiring comprehensive models. Synthetic jets are useful for aerodynamic and flight control.

Adaptive feedback control techniques for actuator nonlinearity compensation have been studied for the last several decades to achieve desired control objectives in numerous critical applications. However, the parametrized actuator nonlinearity model structure is static, and only the parameters vary. An advantage is that the adaptive inverse compensator can be built upon the nominal guidance and control system without the need for a complete controller redesign. Many actuator models, though, have a state-dependent model characteristic as has been seen in synthetic jets. Adaptive control theorists attempt to utilize constrained state-dependent switching algorithms to achieve control objectives when the actuator model is highly uncertain, and a single continuous model is not feasible. However, when the actuator model is dependent on the angle of attack (a state of the aircraft dynamics), switching would not achieve a corresponding mathematical model of the actual actuator physical characteristics. Such state-dependent actuator nonlinearity models are present in several control applications such as electro-hydraulic actuator systems, where in addition to the electrical input signal, the actuator model is also dependent on the sensed position of the valve control element.

This book presents design conditions, control designs, stability analysis, and performance evaluations of adaptive schemes for linear and nonlinear aircraft dynamic systems with simulation results to illustrate the effectiveness. Chapter 1 introduces linear and nonlinear actuators, and specifically, a type of nonlinear actuators called synthetic jet actuators which are useful in flight control and active cooling of microelectronics. Chapter 2 presents the physical characteristics of a synthetic jet actuator, followed by the mathematical model of the actuator and adaptive compensation of the uncertainties. We also present the control of next-generation aircraft with synthetic jet actuation-based virtual aerodynamic wing shaping. The adaptive inverse formulation helps cancel the jet nonlinearities. A robust state feedback law controls the aircraft dynamics with modeling errors and parametric uncertainties. Parameter projection-based update laws ensure a stable system with desired signal boundedness.

Chapter 3 extends the adaptive scheme to a nonlinear aircraft. Adaptive inversion is employed to compensate for the jet nonlinearities at low angles of attack. A nonlinear state feedback law controls the aircraft dynamics with other states providing the intermediate control laws, apart from the lift forces. Chapter 4 presents an adaptive compensation scheme at high angles of attack. We deploy a linearly parametrized function to approximate a nonlinearly parametrized actuator model. Adaptive inverse arrays are robust in compensating the uncertainties both in the true jet nonlinearity model and also the linearly approximated model. A nonlinear state feedback law controls the aircraft dynamics.

Chapter 5 develops a unified adaptation scheme to control state-dependent actuator nonlinearities in a class of nonlinear systems using two approximators. A linearly parametrized function approximates the nonlinearly parametrized actuator model. Adaptive compensation takes place through a second approximator that is used as a feedforward function. Chapter 6 develops an improved high-order parametrization scheme using twin neural networks. One of the neural networks represents the signal-dependent actuator nonlinearities, and the other represents the adaptive compensator. Higher order parametrization improves the accuracy of the approximation presented as evidenced by the application to a nonlinear pitch plane dynamic model. Chapter 7 extends this adaptive control technology for control of state-dependent actuator nonlinearities to a six degrees of freedom aircraft dynamical system for a wide range of angles of attack using two Taylor series approximators.

The authors would like to thank Dr. Douglas Smith who is a Former Assistant Professor, University of Wyoming, and presently with the Office of Naval Research, Govt. of the USA. Dr. Smith's wind tunnel tests were instrumental in the development of synthetic jet actuator models at low and high angles of attack. Funding received from US Air Force for a Project titled, "Robust Adaptive Control of Innovative Actuation Devices for Aircraft," 06-HYP-06-0049, and another under a NASA STTR grant titled, "Active Flow Control with Adaptive Design Techniques for Improved Aircraft Safety," proposal number: 05 T2.02-9831, are duly acknowledged.

Ahmedabad, Gujarat, India — Dipankar Deb
Charlottesville, Virginia, USA — Jason Burkholder
Charlottesville, Virginia, USA — Gang Tao

The original version of the book was revised: Affiliation of the author Dipankar Deb has now been corrected. The correction to the book is available at https://doi.org/10.1007/978-981-16-4161-9_8

Contents

Acronyms and Symbols

AIC	Adaptive Inverse Compensation
AoA	Angles of Attack
BANTAM	Barron Associates Nonlinear Tailless Aircraft Model
IMU	Inertial Measurement Unit
LES	Large Eddy Simulations
NN	Neural Network
R_n	Reynolds Number
SJ	Synthetic Jet
SJA	Synthetic Jet Actuator
SJAA	Synthetic Jet Actuator Arrays

List of Figures

List of Tables

Chapter 1
Introduction

In this book, our focus is on SJAs, a type of nonlinear actuators that are driven by a sinusoidal electrical signal, but which produce a train of vortices, thereby favorably modifying the ambient fluid. More details about the physics of SJAs are in Chap. 2. It is therefore important to briefly introduce the difference between linear and nonlinear actuators.

1.1 Linear and Nonlinear Actuators

A linear actuator typically converts energy into linear motion using diverse types of energy sources like pressure, heat and electricity. Considering the output and power levels, and physical dimensions, the desired objective determines the specific type of actuator used in a given application. Some examples of linear actuators include DVD player trays and certain hydraulic lifts. Linear actuators pose limited control challenge because the input-output relationship is generally a pre-defined and proportional one.

Nonlinear actuators, on the other hand, convert the energy into nonlinear motion, that is, one nonlinear form into another. These transducers have proven highly effective in several industrial, automotive, defense, aerospace and aeronautical applications due to their dual capability for sensing and actuating, the magnitude of generated forces, and their broadband nature. At high drive levels, all of these materials exhibit constitutive nonlinearities, and varying hysteresis degrees. The goal is to develop mathematical models and control algorithms which, in conjunction with drive electronics, compensate for the nonlinearities inherent to these materials. The strategy in the modeling component focuses on utilizing energy mechanisms to develop unified models that are appropriate for each of the smart material compounds. Inverse functional compensators based on these models when augmented with feedback,

D. Deb et al., *Adaptive Compensation of Nonlinear Actuators for Flight Control Applications*, Studies in Systems, Decision and Control 386,
https://doi.org/10.1007/978-981-16-4161-9_1

feedforward, adaptive and robust control laws, provide the control algorithms necessary to achieve linear transducer dynamics. Such developments advance the state of smart material control systems and extend the capabilities of currently available actuators for high performance applications by facilitating the construction of actuators with the full stroke and force configurations of the nonlinear constituent smart materials but exhibit linear behavior throughout their drive range. To illustrate the inherent nonlinearities in these nonlinear actuators, we consider SJAs as a case study and focus entire on this specific type of nonlinear actuators.

1.2 Applications of Synthetic Jet Actuators

In the outset, we present a brief review of the different applications of SJAs. We include a review of existing literature involving SJAs and related to flight control herewith. Even though researches have looked at the flow rearrangement, separation and virtual shaping of the SJs in detail over the past several decades, the heat transfer capability is comparatively less studied in literature and has only picked up over the last decade or so. As a result, naturally, the control aspects of SJs in the context of flight control are more mature as compared to the electronic cooling related control aspects when using SJs. In this book, we have given a detailed account of the control challenges and how to resolve them in the context of flight control.

Apart from thermal management of electronics and aircraft flight control, SJAs apply to wind turbine blades [1, 2] and as a feasible propulsion mechanism for underwater vehicles [3, 4] for similar purposes of flow control, although an extensive work in these applications have not taken place till date. The objective of applying SJAs to wind turbine blades is to reduce the amplitude of blade structural vibrations using active flow control techniques by selectively reattaching the flow along the blade span, thereby manipulating the aerodynamic load along the span. Geng et al. propose an underwater steering method using two SJAs with nozzles aligned in the same direction, and appropriate phase lag adjustment provides lateral and yaw motion of the underwater vehicle [5].

The addressing of the feedback control challenges in the context of thermal management is more nascent in comparison, and much work is due in this direction. We deal with the physical description of the SJAs, modeling in terms of uncertain parameters and appropriate control of the same, in the next chapter.

1.2.1 Synthetic Jet Actuators in Active Cooling

There is a rapid increase in power consumption in the newer processors, and the circuit densities are also seeing a sharp rise. Apart from these challenges, there is also pressure to ensure that the size of such devices keeps reducing. There is a need for path-breaking progress in low-power systems for the conventional fan-driven cooling is no longer enough. Overheating is the primary cause for the failure of internal components in electronics, and the operative reason is ineffective heat dissipation. The simple structure, lightweight and low cost of SJs make them a promising factor in microelectronic cooling, especially, because the need for heat dissipation is even more acute in narrow passages in such systems.

Simple cooling techniques like natural convection, heat sinks, and fans used to maintain the operating temperature within acceptable limits are no longer suited for next-generation applications which require more robust and efficient ways of cooling. The increasing deployment of VLSI technologies has steadily increased the heat dissipation levels in microelectronic devices to higher and higher levels. The deployment of SJAs for enhancement of heat transfer rate is a prominent methodology for cooling miniature and narrow aperture electronics devices. As compared to conventional jets, no additional source of fluid is required to form the jet; therefore no complex flow system is required and is thus easy to handle and reduces the cost of the thermal management system. Arshad et al. have recently reviewed several factors such as the effect of actuation and geometrical and fluid parameters of SJAs on fluid flow and heat transfer characteristics.

Devoid of any frictional parts, an SJ module replicates a tiny stereo speaker with a piezoelectric diaphragm mounted cavity containing one or more orifices. Requiring lesser airflow than typical fans, SJs suck surrounding air and expel air in and out of the chamber creates pulsating jets applicable at the specific locations needing cooling whenever needed. Yadav et al. [6] and Chaudhari et al. [7] have found that the multi-orifice SJs offers several advantages over a single orifice SJ like increased rate of heat with lesser surface spacing.

The deployment of SJs for heat transfer is promising in many applications like LED cooling, turbine blades, and micro-electronics. In recent times, Light-emitting diodes (LED) have become more reliable, and the performance of LED lights have increased multi-fold with the usage of SJ cooling device [8]. The SJ solutions can also cool different types of electronics and battery heat management [9], and also facilitate air cooling of data centres [10]. Campbell et al. [11] found that the notebook processor temperature reduces by about 26% when compared to free convection. Jagannatha et al. identified that the SJ module is superior to pure convection by at least 20 times in electronic cooling applications and operates sans additional fluid circuits [12]. The SJ heat transfer performance can be further improved by a suitable cavity shape choice [13].

The cooling effectiveness of SJs also varies as the distance separating the nozzle and the plate changes [14, 15]. Valiorgue et al. [16] found that the heat transfer capability does not have a direct relationship with stroke length beyond a critical value. Chaudhari et al. [17] have explore heat exchange capabilities of SJs of different orifice shapes and found the square orifice is most efficient with the targeted flat surface in the far-field and the elliptical one is best in the near-field. Fanning et al. [18] and Greco et al. [19] study the flow characteristics and heat exchange of two interacting and adjacently placed SJs and find that when operating out-of-phase the SJs with sufficient gap behave as if they are two single SJs. The interaction happens when SJs are either close to each other or in-phase. Li et al. [20] have shown that cooling performance increases with increased number of SJs. However, power consumption is a limiting factor in keeping on increasing that number, and an optimal number should is experimentally determined. Jet placement and jet-to-target distance are also essential factors. Timchenko et al. [21] found through numerical studies that usage of an SJ in a 200 μm micro-channel increases heat transfer by more than 60%. Many researchers have found through experiments that SJs can suppress thermally induced flow instabilities in micro-channels.

A detailed study of relevant literature indicates that most studies for flow and thermal characteristics of the SJ are limited to numerical modeling. This limitation is one of the reasons why we have looked extensively into different aspects of flow control while using SJs for virtual shaping for flight control in this book. On similar lines, there is a need to study, for instance, the control challenges in the coordinated cooling while using multiple SJs in a constrained environment. We need an increased emphasis on conducting experiments for adequate validation. Also, more work needs to aim at finding the optimum parameters related to the SJs.

1.2.2 Synthetic Jet Actuators for Flight Control

Flow control aims at improving aircraft performance by locally modifying the flow. For such purposes, there are mechanical and fluidic technologies, the latter presenting the brightest potential. It usually deals with blowing air locally at high speed, through slots or holes. The blowing phase can be continuous or pulsed. In both cases, the energy required is a constraint because the overall energy balance is not always favourable. SJ technology is an alternative to continuous or pulsed jets. It consists of creating high-frequency alternating suction and blow phases. The advantage is that no additional air source is needed since the air gets sucked from the main flow (zero-net-mass-flux). It is, therefore, a technology with high application potential, and several studies have shown the contribution of SJs with significant performance gains. A typical SJA used for flow control applications is shown in Fig. 1.1. The wires visible in the figure are used to provide electrical power to the SJA.

Fig. 1.1 A typical Synthetic Jet Actuator for active control action

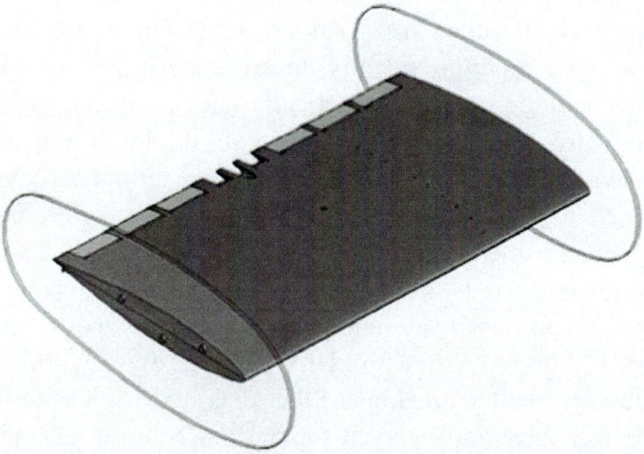

Fig. 1.2 A typical design with SJA in the trailing edge for virtual airfoil shaping

Studies show that SJAs are efficient flow-separation controlling devices [22, 23] and can also virtually change the shape of airfoils through flow reattachment for desired performance [24]. Using a symmetric airfoil, Amitay et al. [25], and Crook and Wood [26] explored SJAs under different testing conditions. The control of SJs prevents boundary layer separation at high AoA. Such control action provides a significant gain in the aerodynamic lift force and dramatic reductions in the drag force. The third component of the aerodynamic force, the side force is not considered in this study. The lift and drag forces are dependent on the AoA (one of the states of the dynamic system) of the aircraft. Wind tunnel tests reveal that the angle of attack is also a factor in the actuator nonlinearities. Therefore, a new and improved state-dependent adaptive compensation control methodology is one of the objectives of this book.

Apart from the aerodynamic and the gravitation forces, the aircraft also experiences the propulsive force (generated by the propulsion section) referred to as the thrust force. This force acts as a control for other aircraft states which cannot be accounted for the aerodynamic forces directly. The pressure distribution profile over the airfoils studied at different AoA provides insight into the effect of flow reattachment reattached on the applied control signals. Flow control studies by Washburn and Amitay [27] demonstrate that apart from flow separation control at high AoA, SJAs are effective in controlling vehicle roll and pitch, and use the modulation frequency of the SJA to control vehicle attitude.

Virtual airfoil shaping is another aspect dealt with extensively in the literature. A typical design with SJA in the trailing edge for virtual airfoil shaping is shown in Fig. 1.2.

First deployed on a cylindrical model, Honohan et al. [28], show that SJAs displace streamlines and create recirculating regions in the vicinity of the orifice. To recreate a similar condition, Chatlynne et al. [29] utilized SJAs along with a small obstacle to generate separated flows and a stationary recirculating domain. Chen and Beeler [30] noticed an increased surface pressure at low AoA actuation of SJs over a two-dimensional NACA 0015 model in a subsonic wind tunnel. Pack and Seifert [31] experimentally surmise that for the desired outcome from SJAs around four vortices are needed over an airfoil surface at any given time. Wind tunnel tests encompassing various conditions to fully describe the impact of SJ arrays over a diverse range of operation modes is practically hard to accomplish, and so an AIC technique is attractive. Such a method allows robust, closed-loop aircraft control using SJ arrays overcoming prior limitations to achieve virtual aerodynamic shaping.

SJ-based separation control is extensively explored through numerical simulations and experimental studies over different variants of airfoils [32]. Avdis et al. found around 30% reduction in the size of the separated zone with SJA control both in LES and experimental validation on a turbulent flow over on a flat plate at $R_n = 9.4 \times 10^5$ [33]. Working with a similar simulation framework at $R_n = 4 \times 10^3$, Suzuki et al. observe that the breaking of the separated vortices results in improved aerodynamic performance and delayed separation. Naim et al. infer through experiments on a circular cylinder that the SJA position, forcing frequency and direction vis-a-vis the cross-flow are the main factors [34]. Through an LES study, Dandois et al. find that with a rounded ramp, the turbulent boundary layer flow at $R_n = 2.8 \times 10^4$ suppresses the separation zone at low-frequencies, but makes it pronounced at higher frequencies [35].

The implementation of the SJ generally involves a few fixed attributes, and the rest vary within known bounds to surmise the possible control outcomes and associated correlation among those parameters. The control effects and aerodynamic performance are dependent on the SJA shape and the arrangement of the SJAA. The SJA forcing frequency and the mean flow receptivity are the main factors needing optimization for improved separation control. Kotapati et al. notice ineffective forcing phenomena with the merging of Kelvin-Helmholtz vortices over a flat plate and the leading edge in the form of an ellipsoid at high frequencies [36]. Chunzhi et al. have used a NACA 0015 model to ensure a robust roll angle stability using SJAs at 18° AoA while taking the damping effect in roll into account [37]. In a recent work involving wind tunnel tests [38], Hasegawa and Obayashi find that with SJAs on a NACA0012 airfoil the maximum lift coefficient increases by 11% and stall is delayed by 4° vis-a-vis the no SJA control scenario. In a recent work, [39] find that at AoA larger than that which provides lift maximization, SJA actuation ensures reattachment of the separated flow thereby increasing aerodynamic efficiency by 124% and drag reduction by around 45%.

Recent developments in electrically controlled SJAs also involve the chamber walls which work synergistically to reduce or enlarge the volume of the jet actuator chamber in three dimensions simultaneously. The jet velocity and mass flow rate for the actuators are from several times up to a couple of orders magnitude higher than conventional piezoelectric actuators. The actuator is designed either in the horizontal or vertical position and also as a curved chamber instead of the conventionally used cylindrical chamber, to reduce the dead chamber volume and increase the SJA efficiency. Another application is a SJ humidifier which is constructed form a surface SJA. The droplets of water generated are from the atomization of fluid sucked into the orifice of the SJA and the cone formed is dependent on the orifice and chamber geometry. We would also like to point out that the designs developed and demonstrated using synthetic jets are widely applicable to many other closely related types of jets. The control approaches are also applicable to other nonlinear, possibly distributed, flow control actuators. Examples include plasma jets, active blowing jets, and morphing actuators.

References

1. Deb, D., Sonowal, S.: Synthetic jet actuator based adaptive neural network control of nonlinear fixed pitch wind turbine blades. In: 2013 IEEE International Conference On Control Applications (CCA) (2013). https://doi.org/10.1109/cca.2013.6662759
2. Vries, H., Weide, E., Hoeijmakers, H.: Synthetic jet actuation for load control. J. Phys.: Conf. Ser. **555** (2014). https://doi.org/10.1088/1742-6596/555/1/012026
3. Mohseni, K.: Zero-mass pulsatile jets for unmanned underwater vehicle maneuvering. In: AIAA 3rd "Unmanned Unlimited" Technical Conference, Workshop And Exhibit (2004). https://doi.org/10.2514/6.2004-6386
4. Thomas, A., Milano, M., G'Sell, M., Fischer, K., Burdick, J.: Synthetic jet propulsion for small underwater vehicles. In: Proceedings of the 2005 IEEE International Conference on Robotics and Automation. https://doi.org/10.1109/robot.2005.1570116
5. Geng, L., Hu, Z., Lin, Y.: Hydrodynamic characteristic of synthetic jet steered underwater vehicle. Appl. Ocean Res. **70**, 1–13 (2018). https://doi.org/10.1016/j.apor.2017.11.005
6. Yadav, H., Joshi, A., Chaudhari, M., Agrawal, A.: An experimental study of a multi-orifice synthetic jet with application to cooling of compact devices. AIP Adv. **9**(12) (2019). https://doi.org/10.1063/1.5128776
7. Chaudhari, M., Puranik, B., Agrawal, A.: Multiple orifice synthetic jet for improvement in impingement heat transfer. Int. J. Heat Mass Transf. **54**(9–10), 2056–2065 (2011). https://doi.org/10.1016/j.ijheatmasstransfer.2010.12.023
8. Grimm, D.: Advanced Synjet Cooler Design for LED Light Modules, US 8299691 B2, US Patent
9. Glezer, A., Mahalingam, R., Heffington, S.: Synthetic jet heat pipe thermal management system, US 7607470 B2, US Patent (2009)
10. Garimella, S., Persoons, T., Weibel, J., Gektin, V.: Electronics thermal management in information and communications technologies: challenges and future directions. IEEE Trans. Compon. Packag. Manuf. Technol. **7**(8), 1191–1205 (2017). https://doi.org/10.1109/tcpmt.2016.2603600
11. Campbell, S., Black, W., Glezer, A., Hartley, J.: Thermal management of a laptop computer with synthetic air microjets. Itherm'98. In: Sixth Intersociety Conference on Thermal and Thermomechanical Phenomena in Electronic Systems (Cat. No.98CH36208). https://doi.org/10.1109/itherm.1998.689518
12. Jagannatha, D., Narayanaswamy, R., Chandratilleke, T.: Analysis of a synthetic jet-based electronic cooling module. Numer. Heat Transf. Part A: Appl. **56**(3), 211–229 (2009). https://doi.org/10.1080/10407780903163702
13. Bhapkar, U., Srivastava, A., Agrawal, A.: Proper cavity shape can mitigate confinement effect in synthetic jet impingement cooling. Exp. Therm. Fluid Sci. **68**, 392–401 (2015). https://doi.org/10.1016/j.expthermflusci.2015.05.006
14. Persoons, T., McGuinn, A., Murray, D.: A general correlation for the stagnation point Nusselt number of an axisymmetric impinging synthetic jet. Int. J. Heat Mass Transfer. **54**(17–18), 3900–3908 (2011). https://doi.org/10.1016/j.ijheatmasstransfer.2011.04.037
15. Greco, C., Paolillo, G., Ianiro, A., Cardone, G., de Luca, L.: Effects of the stroke length and nozzle-to-plate distance on synthetic jet impingement heat transfer. Int. J. Heat Mass Transf. **117**, 1019–1031 (2018). https://doi.org/10.1016/j.ijheatmasstransfer.2017.09.118

16. Valiorgue, P., Persoons, T., McGuinn, A., Murray, D.: Heat transfer mechanisms in an impinging synthetic jet for a small jet-to-surface spacing. Exp. Therm. Fluid Sci. **33**(4), 597–603 (2009). https://doi.org/10.1016/j.expthermflusci.2008.12.006
17. Chaudhari, M., Puranik, B., Agrawal, A.: Effect of orifice shape in synthetic jet based impingement cooling. Exp. Therm Fluid Sci. **34**(2), 246–256 (2010). https://doi.org/10.1016/j.expthermflusci.2009.11.001
18. Fanning, E., Persoons, T., Murray, D.: Heat transfer and flow characteristics of a pair of adjacent impinging synthetic jets. Int. J. Heat Fluid Flow **54**, 153–166 (2015). https://doi.org/10.1016/j.ijheatfluidflow.2015.05.005
19. Greco, C., Castrillo, G., Crispo, C., Astarita, T., Cardone, G.: Investigation of impinging single and twin circular synthetic jets flow field. Exp. Therm. Fluid Sci. **74**, 354–367 (2016). https://doi.org/10.1016/j.expthermflusci.2015.12.019
20. Li, R., Gerstler, W., Arik, M., Vanderploeg, B.: Heat transfer impact of synthetic jets for air-cooled array of fins. J. Heat Transf. **138**(2) (2015). https://doi.org/10.1115/1.4031647
21. Timchenko, V., Reizes, J., Leonardi, E.: A numerical study of enhanced micro-channel cooling using a synthetic jet actuator. In: 15th Australasian Fluid Mechanics Conference, Sydney (2014)
22. Amitay, M., Parekh, D., Pitt, D., Kibens, V., Glezer, A.: Control of internal flow separation using synthetic jet actuators. In: 38th Aerospace Sciences Meeting and Exhibit (2000). https://doi.org/10.2514/6.2000-903
23. Smith, D., Amitay, M., Kibens, V., Parekh, D., Glezer, A.: Modification of lifting body aerodynamics using synthetic jet actuators. In: 36th AIAA Aerospace Sciences Meeting and Exhibit (1998). https://doi.org/10.2514/6.1998-209
24. Amitay, M., Parekh, D., Smith, D., Kibens, V., Glezer, A.: Aerodynamic flow control over an unconventional airfoil using synthetic jet actuators. AIAA J. **39**, 361–370 (2001). https://doi.org/10.2514/3.14740
25. Amitay, M., Horvath, M., Michaux, M., Glezer, A.: Virtual aerodynamic shape modification at low angles of attack using synthetic jet actuators. In: 15Th AIAA Computational Fluid Dynamics Conference (2001). https://doi.org/10.2514/6.2001-2975
26. Crook, A., Wood, N.: Measurements and visualisations of synthetic jets. In: 39th Aerospace Sciences Meeting and Exhibit (2001). https://doi.org/10.2514/6.2001-145
27. Washburn, A., Amitay, M.: Active flow control on the STINGRAY UAV: Physical Mechanisms. In: 42nd AIAA Aerospace Sciences Meeting and Exhibit (2004). https://doi.org/10.2514/6.2004-745
28. Honohan, A., Amitay, M., Glezer, A.: Aerodynamic control using synthetic jets. In: Fluids 2000 Conference and Exhibit (2000). https://doi.org/10.2514/6.2000-2401
29. Chatlynne, E., Rumigny, N., Amitay, M., Glezer, A.: Virtual aero-shaping of a Clark-Y airfoil using synthetic jet actuators. In: 39th Aerospace Sciences Meeting and Exhibit (2001). https://doi.org/10.2514/6.2001-732
30. Chen, F., Beeler, G.: Virtual shaping of a two-dimensional NACA 0015 airfoil using synthetic jet actuator. In: 1St Flow Control Conference (2002). https://doi.org/10.2514/6.2002-3273
31. Pack, L., Seifert, A.: Dynamics of active separation control at high reynolds numbers. AIAA Paper No. 2000-0409, 38th Aerospace Sciences Meeting & Exhibit, Reno, NV (2000)
32. Zhang, W., Samtaney, R.: A direct numerical simulation investigation of the synthetic jet frequency effects on separation control of low-Re flow past an airfoil. Phys. Fluids **27** (2015). https://doi.org/10.1063/1.4919599
33. Avdis, A., Lardeau, S., Leschziner, M.: Large eddy simulation of separated flow over a two-dimensional hump with and without control by means of a synthetic slot-jet. Flow Turbul. Combust. **83**(3), 343–370 (2009). https://doi.org/10.1007/s10494-009-9218-y
34. Naim, A., Greenblatt, D., Seifert, A., Wygnanski, I.: Active control of a circular cylinder flow at transitional Reynolds numbers. Flow Turbul. Combust. **78**(3–4), 383–407 (2007). https://doi.org/10.1007/s10494-007-9068-4
35. Dandois, J., Garnier, E., Sagaut, P.: Numerical simulation of active separation control by a synthetic jet. J. Fluid Mech. **574**, 25–58 (2007). https://doi.org/10.1017/s0022112006003995

36. Kotapati, R., Mittal, R., Cattafesta, L.: Numerical experiments in synthetic jet based separation control. In: 44th AIAA Aerospace Sciences Meeting and Exhibit (2006). https://doi.org/10.2514/6.2006-320
37. Chunzhi, L., Jianying, Y., Yifan, L.: The robust roll control using only synthetic jet actuators in high-angle-of-attack flight. In: 2015 34th Chinese Control Conference (CCC) (2015). https://doi.org/10.1109/chicc.2015.7260078
38. Hasegawa, H., Obayashi, S.: Active stall control system on NACA0012 by using synthetic jet actuator. J. Flow Control Meas. Vis. **07**(01), 61–72 (2019). https://doi.org/10.4236/jfcmv.2019.71005
39. Rodriguez, I., Lehmkuhl, O., Borrell, R.: Effects of the actuation on the boundary layer of an Airfoil at Reynolds Number Re = 60000. Flow Turbul. Combust. **105**, 607–626 (2020). https://doi.org/10.1007/s10494-020-00160-y

Chapter 2
Synthetic Jet Actuators and Arrays: Modeling and Control

As will be evident again and again in this book, AoA has a direct influence on the actuator nonlinearities. Figure 2.1 shows an airfoil at two different angles; the yellow colored airfoil represents a situation under low AoA, and the red color one indicates a situation under high AoA. The characteristics of the SJA under these situations is different because the ambient flow separates from the surface closer to the leading edge as the AoA increases. As a result the affect of SJAs on the ambient flow is different in the two cases. The resultant lift and drag forces are also distinct.

As discussed in the first chapter, an extensive experimental or numerical study encompassing all kinds of conditions that affect SJAs and their possible control of aircraft wings, is not feasible. It is therefore, critical to know the characteristics of SJAs enough to create a mathematical model that is usable for different conditions. Therefore, in this chapter, we first give a physical description of an SJA and describe the virtual shaping of an airfoil at low AoA using SJAs [4]. We formulate actuator mathematical models (and the parametrized equivalent) of the actuators and a feasible distribution of the actuators on the airfoil. We also develop adaptive compensation methods for actuator nonlinearities.

2.1 Physical Characteristics

An SJA is an actuator that creates a fluid flow of zero mass over a full cycle, but the momentum is non-zero beyond a nozzle on the other side of a piezo-electrically driven diaphragm which imposes an oscillating pressure gradient, and as a result, the fluid enters and exits the orifice alternately. The actuator jet velocity is roughly inversely proportional to both the orifice and cavity depth.

The velocity difference between the exiting fluid and the laminar flow causes the air blown out to create a shear layer between the two fluids. During suction, the fluid

D. Deb et al., *Adaptive Compensation of Nonlinear Actuators for Flight Control Applications*, Studies in Systems, Decision and Control 386,
https://doi.org/10.1007/978-981-16-4161-9_2

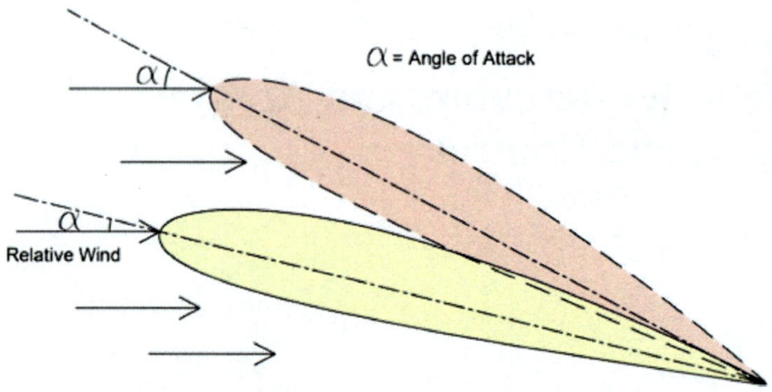

Fig. 2.1 Angle of attack (Low and High)

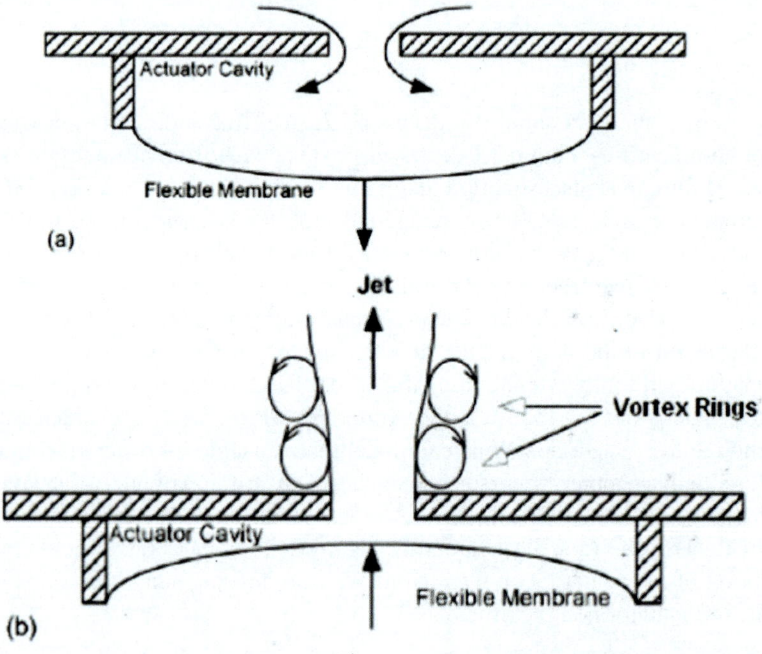

Fig. 2.2 SJA: **a** suction and **b** exhaust stage of operation

enters the actuator cavity, and the vortex pair generated during the blowing period just before that suction period, moves away from the nozzle and does not revert back. Periodic repetition of this process with self-induced velocity leads to a train of vortices to forms the SJ, as shown in Fig. 2.2. The flowfield established in the vicinity and downstream of the nozzle is by and large undisturbed and the train of vortices retain across multiple cycles leading to the maintenance of the SJ. The SJA can bend, or vector, a jet stream from another jet actuator. The air flow inside the chamber is compressible at frequencies close to and beyond Helmholtz frequency.

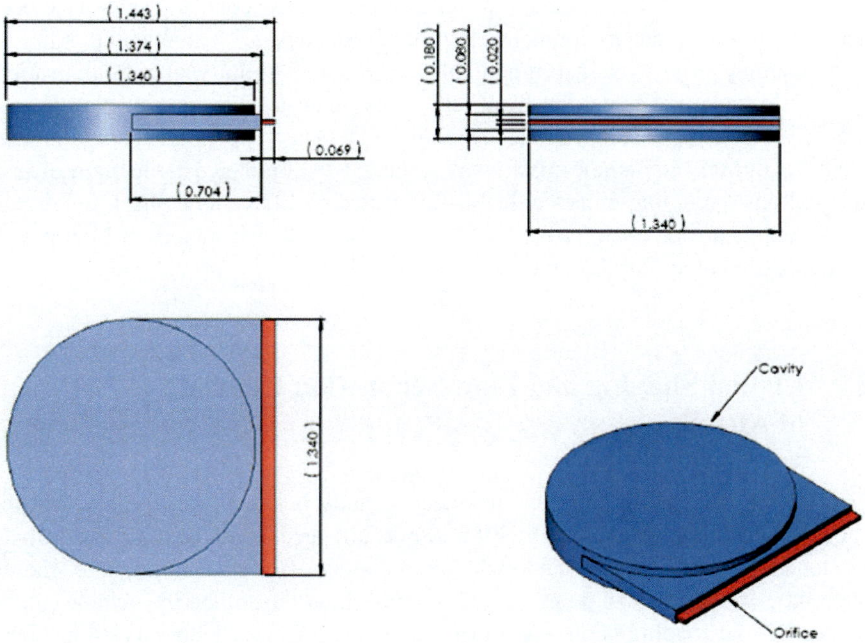

Fig. 2.3 Actuator cavity (blue) and orifice (red) dimensions (in inches)

The evolution of jet for different actuation frequencies observed using a high-speed camera reveals that the rate of discrete vortex ring formation is proportional to the actuation frequency [8]. The vortex rings formed at lower frequencies (10 Hz) move far downstream before being dissipated into turbulence. The geometry of the SJAs needs to accommodate high-pressure nozzles and produce droplets with average sizes of 92-116 μm at maximum rates of 19–28 μL/s [9]. Feero et al. presented an experimental validation of the dependence of SJA performance on the cavity shapes, with the parameters like volume of cavity, and length and diameter of orifice same [10].

We employ a piezo-electrically operated resonant cavity. An SJA (in terms of cavity and orifice dimensions) found to be suitable for usage on aircraft for virtual shaping at low AoA, is shown in Fig. 2.3. Currently, most of the SJAs utilize piezoelectric materials to produce the desired jet flow. However, in some applications, piezoelectric materials may not give a large force to induce an adequate jet flow.

There is a need for an innovative new SJA design that promises to overcome many of the limitations of traditional SJAs for a much broader dynamic range, improved manufacturing uniformity and scalability, and vastly improved environmental robustness. A series of environmental tests and subsequent wind tunnel tests, including an existing adaptive feedback controller, are used to assess peak performance and environmental robustness and objectively compare the design approaches in terms of reliability, cost, weight, etc. Actuator failure modes and failure probabilities can be statistically modeled. Although promising active flow control results have been achieved in controlled environments, traditional SJA designs feature inherent limitations that may prove to be insurmountable in a practical military aircraft application. For SJs to make the transition from laboratory scale tests to aircraft platforms, a robust SJA must be developed that can withstand the harsh operational environments typical of military air vehicle missions. A suitable SJA could give rise to many diverse applications, including enhancement of aircraft high lift systems, back-up aircraft flight control power for improved safety, and many other instances of high-payoff lift enhancement and/or drag reduction.

2.2 Virtual Shaping and Flow Separation Control of Airfoils

Flap lies on the trailing edge of the wing, usually between the fuselage and the ailerons, and aim to generate more lift at slower airspeed, enabling the airplane to fly at a much-reduced speed with a lower risk of stalling. However, if the same amount of aerodynamic force can be generated without these flaps, then the vehicle can be lighter. Virtual shaping is one such process, and takes place with varying levels of effectiveness at all AoA. However, it is most effective at low AoA when applied at the airfoil trailing edge due to minimal flow separation. With virtual shaping, the actuators dynamically operate through an intelligent controller to for manoeuvrability of the air vehicle.

A stationary re-circulation flow domain is created by placing a SJA downstream of the obstruction, thereby modifying the airfoil shape virtually as shown in Fig. 2.4. SJ flow control action causes reattachment of a separated flow (also controls the degree of flow separation) to ensure this virtual shape change and the extent of such

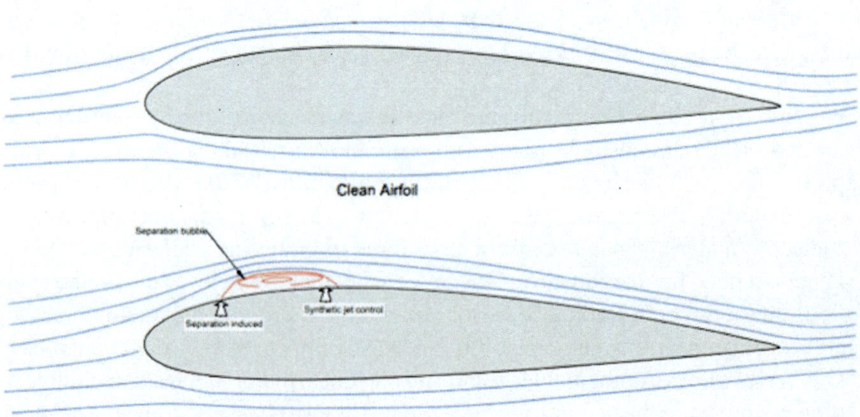

Fig. 2.4 Illustration of modified streamlines with virtual shaping

a change. The bubble is shown exaggerated to provide clarity. The change in the distributed pressure in the vicinity of the bubble decreases the drag of the airfoil with only a minimal change in the lift.

When separation takes place, the adjacent flow is unsteady, and the resulting turbulent flow is more vigorous compared to the attached boundary layers. At the separation point (zero fluid velocity), the originating streamline moves away from the surface. The flow downstream is separated and generally moves upstream. Pressure rise (velocity drop) results in the laminar flow moving away, causing turbulence. Post separation, when this flow in the surface vicinity is again steady and smooth, it is called the reattached flow. The shape of this reattached section is possible to control by actuating the SJs and modulating the control signal.

Eliminating (or reducing) the mechanical deflecting surfaces by replacing with SJAs could result in a reduction in the cost, weight, and vehicle complexity. The number of SJs and their optimal distribution is critical for adequate virtual shaping is determined experimentally by wind tunnel tests. The separation of the boundary layer at high AoA causes an uncontrolled flow around the airfoil containing large vortices representing separated streamlines which induce a drastic decrease in lift and wing stalls in the absence of control. By introducing SJAs at an appropriate point, and exciting the flow at a given frequency, the separated flow gets reattached, thereby creating airfoil virtual shape change. The aircraft stall can get potentially delayed with SJA control at the leading edge is shown in Fig. 2.5.

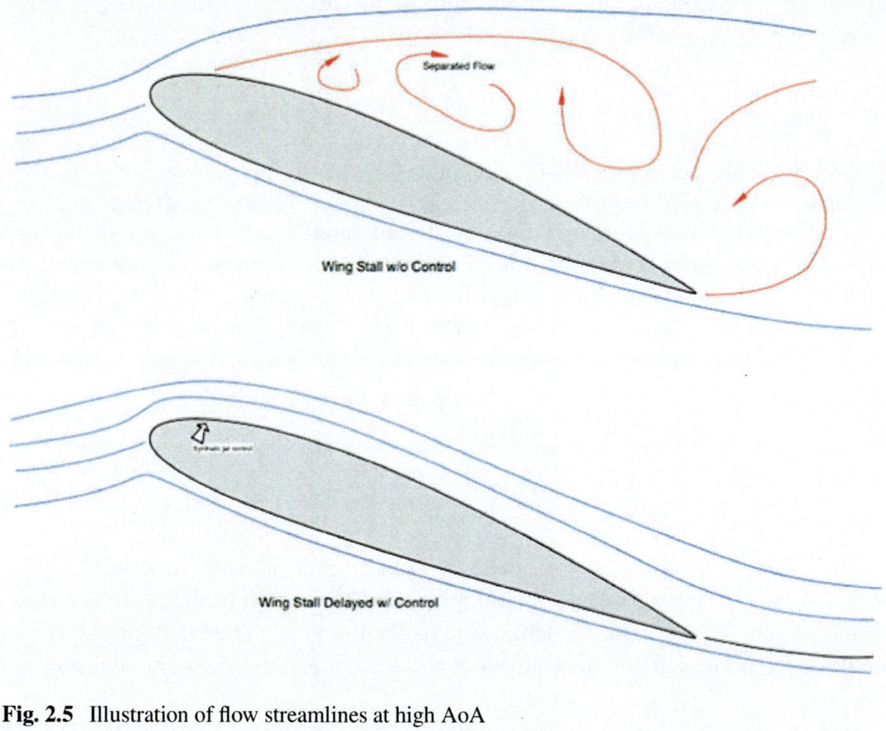

Fig. 2.5 Illustration of flow streamlines at high AoA

Flow separation control on NACA23012 airfoil with SJs demonstrate that it is possible to significantly improve stall characteristics and control the surface performance by a process of resizing of the separation vortices. When the separation point coincides with the jet location, maximum lift is possible, and the non-dimensional frequency is around 1. The separation control effect is proportional to the peak velocity of the jet [11]. Delaying flow separation allows an aircraft to possibly recover from adverse conditions that would otherwise cause loss of control or stall. An aerodynamic model is required to allow the design of controls and to evaluate the capabilities of the aircraft at high AoA.

2.3 Mathematical Models

The optimal non-dimensional frequency F^+ (normalized by chord length and freestream velocity) [1] derived from the Strouhal number is given by

$$F^+ = \frac{1}{u_\infty} c\, f_{act}, \tag{2.1}$$

where f is a constant actuation frequency, c is the airfoil chord length the airfoil, and u_∞ is the freestream velocity.

A parametric model representing a SJA is developed [2, 3] Consider a SJA with an input voltage of amplitude $v(t)$ and equivalent virtual deflection $u_a(t) = C_l(t)$, expressed by a nonlinearity $N(\cdot)$:

$$u_a(t) = C_l(t) = N(v(t)). \tag{2.2}$$

We assume that $u_a(t)$ is the actual input that is not measurable, and all lower bounds parameters of $N(\cdot)$ are known.

The actuator output decreases dramatically and nonlinearly as the operating frequency deviates from resonance. For adequate control authority, the actuator frequency set lower than resonance, is given as f_{mod}. The jet actuation is such that the generated control signal causes a favorable change of the virtual shape change of the airfoil. Under constant air stream density and flow control along the width, and momentum coefficient $C_\mu(t) = \frac{p_4 A_{pp}^2(t)}{U_\infty^2(t)}$, the SJA model at low AoA is

$$C_l(t) = p_1 - \frac{p_2\, p_3}{F^+ C_\mu}. \tag{2.3}$$

The unknown parameters $(p_i, i = 2, 3, 4)$ of certain physical meaning are the positive scalars as per a baseline or a reference actuator model meant to study airfoil virtual shaping. The maximum achievable deflection is p_1. The effect of the SJ on the lift force translates into an equivalent surface deflection. The parameters $p_1 =$

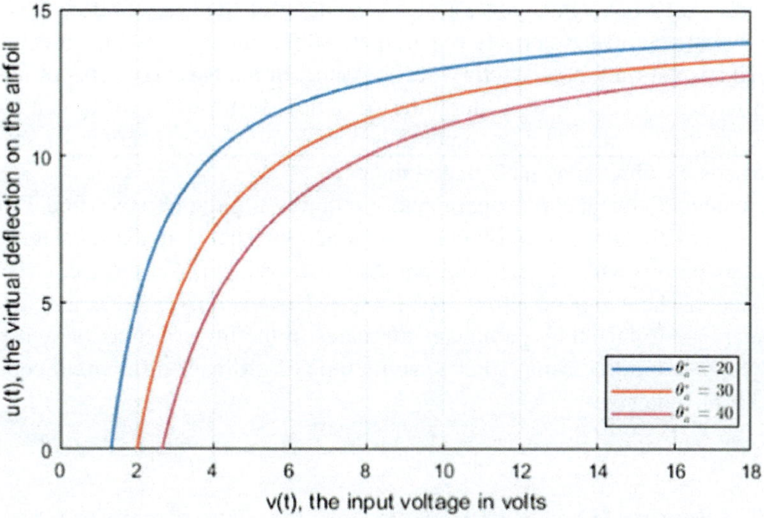

Fig. 2.6 Virtual airfoil deflections with actuation voltage

15, $p_2 = 0.05$, $p_3 = 0.05$, and $p_4 = 5$ are chosen arbitrarily, and $c = 2$ft, $f = 60$Hz, and $u_\infty = 200$ ft/s are also chosen. Utilizing these parameters, we get

$$C_l(t) = 15 - \frac{33.33}{A_{pp}^2(t)}. \tag{2.4}$$

Figure 2.6 describes virtual airfoil deflections against actuator voltage for three values of θ_a^*.

For adaptive compensation, we consider parameters $\theta^* = [\theta_a^*, \theta_b^*]^T \in R^2$ for the nonlinearity function, $N(\cdot)$, expressed as

$$u_a(t) = \theta_b^* - \frac{\theta_a^*}{v(t)} = N(\theta^*; v(t)), \tag{2.5}$$

where the virtual deflection $u_a(t)$ ranges from 0 to 15, and the input voltage $v(t)$ is always positive.

2.4 Adaptive Inverse Compensation and Update Laws

Adaptive control techniques deal with complex systems that have unpredictable parameter deviations and uncertainties. Many control techniques are available in the literature to handle such parametric uncertainties related to SJAs. For instance, Pedroza et al. have proposed a SJA-based nonlinear robust controller capable of suppressing limit cycle oscillations with parametric uncertainties. Adaptive control has been an area of intense interest of researchers for the last half a century or so because most practical systems have parameters that change with time or conditions and need adaptation for satisfactory performance. Some basic concepts of adaptive control have been reviewed in literature [5]. Adaptive inverse control is a specific type of adaptive control that attempts to generate an inverse function for actuators [6] and sensors which may have uncertainties.

An actuator with uncertain parameters requires a parametrized model and an adaptive estimator in the form of an inverse function. That is, under ideal and hypothetical conditions when the actuator parameters are known, the inverse is such that the product of the two parametrizations is unity. However, in practice, the uncertain parameters are reflected by parameter estimates in the inverse function and subsequently updated online using some specific type of ordinary differential equations called adaptive laws.

2.4.1 Adaptive Inverse Function

For this purpose, with the help of a parameter estimation vector $\theta(t) = [\theta_a(t), \theta_b(t)]^T$ to compensate for the uncertain parameter vector θ^*, an appropriate adaptive inverse function is given as

$$v(t) = \widehat{NI}(u_d(t)) = \frac{\theta_a(t)}{\theta_b(t) - u_d(t)}, \tag{2.6}$$

where instead of the actuator output $u(t)$, the desired input u_d(t) is inserted and the error between these two signals is to be computer. The signal $u_d(t)$ is internally provided using state feedback.

The parameters estimate θ should remain in specific regions ($\theta_a, \theta_b > 0$) for the inverse function implementation, and are governed by projection-based adaptive update laws. To ensure that the inverse function We make an assumption regarding the desired control law $u_d(t)$ that

(A1) $u_d(t) < \theta_b(t), \forall t \geq 0$, for non-singularity of the inverse function.

Using the condition $\theta_a(t) > 0$ and assumption (A1), we get that $v(t) \neq 0$ and so (2.6) is singularity-free. From (2.5) and (2.6), we formulate an equation

$$u_a(t) = \frac{\theta_b^* \theta_a(t) - \theta_a^* (\theta_b(t) - u_d(t))}{\theta_a(t)}, \tag{2.7}$$

which is used to find the control error with a re-organization of the terms, given as

$$u_a(t) - u_d(t) = \left(\frac{\theta_b(t) - u_d(t)}{\theta_a(t)}\right) \left(\theta_a(t) - \theta_a^*\right) - \left(\theta_b(t) - \theta_b^*\right), \tag{2.8}$$

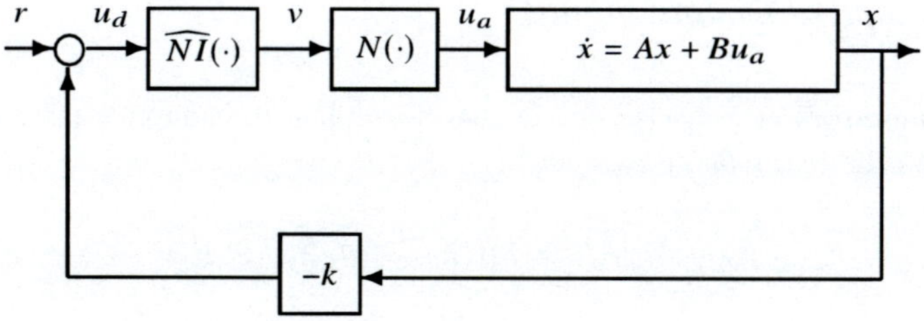

Fig. 2.7 Adaptive inverse control action with state feedback

where $u(t)$ is applied, and $\theta_a(t)$ and $\theta_b(t)$ are updated adaptively. Such a control system is illustrated in Fig. 2.7.

For an input $u_a(t)$, consider a controllable linear time-invariant system given as

$$\dot{x}(t) = Ax(t) + Bu_a(t),\ x(t) \in R^{n\times n},\ u_a(t) \in R, \tag{2.9}$$

where $A \in R^{n\times n}$ and $B \in R^{n\times 1}$ are known and constant.

The parametrized function $N(\theta^*; v(t))$ is given in (2.5). The adaptive inverse $\widehat{NI}(u_d(t))$ for $N(\cdot)$ is defined in (2.6). For a bounded reference signal $r(t)$, the desired control signal is given by

$$u_d(t) = -kx(t) + r(t), \tag{2.10}$$

where $k \in R^{1\times n}$ is a constant gain chosen from linear regulator [21] or pole placement method [24] to ensure that the eigenvalues of $A - Bk$ have negative real parts.

Using (2.8), (2.9), and (2.10), we have

$$\begin{aligned}\dot{x}(t) &= (A - Bk)x(t) + Br(t) - B(\theta_b(t) - \theta_b^*)\\ &\quad + B\left(\frac{\theta_b(t) - u_d(t)}{\theta_a(t)}\right)(\theta_a(t) - \theta_a^*).\end{aligned} \tag{2.11}$$

Considering the structure of the system, it is reasonable to choose a reference system

$$\dot{x}_r(t) = (A - Bk)x_r(t) + Br(t), \tag{2.12}$$

such that for gain k and estimation-based adaptive update laws for $\theta(t)$, boundedness is ensured for all closed-loop signals and $\lim_{t\to\infty}(x(t) - x_r(t)) = 0$. We assume that the lower bounds of θ_i^* are known, i.e., $0 \le \theta_i^a \le \theta_i^*$. Initial estimates $\theta_i(0)$ of θ_i^* ensure that $\theta_i(0) \ge \theta_i^a$.

2.4.2 Adaptive Update Signals

For tracking errors $e(t) = x(t) - x_r(t)$, we formulate the adaptive parameter estimation laws for $\theta_a(t), \theta_b(t)$ as

$$\dot{\theta}_i(t) = g_i(t) + f_i(t),\ i = a, b,\ t \geq 0, \tag{2.13}$$

$$\text{where } g_a(t) = -\gamma_a e^T(t) P B \frac{(\theta_b - u_d(t))}{\theta_a},\quad g_b(t) = \gamma_b e^T(t) P B,$$

with adaptation gains $\gamma_i > 0, i = a, b$ and $P \in R^{n\times n}$, $P = P^T > 0$ satisfying

$$P A_r + A_r^T P = -Q,\ A_r = A - Bk \tag{2.14}$$

for a constant $Q \in R^{n\times n}$, $Q^T = Q > 0$,

$$f_i(t) = \begin{cases} 0 & \text{if } \theta_i(t) > \theta_i^a,\ \text{ or} \\ & \text{if } \theta_i(t) = \theta_i^a \text{ and } g_i(t) \geq 0 \\ -g_i(t) & \text{otherwise.} \end{cases} \tag{2.15}$$

2.4.3 Stability Analysis

Next, we ascertain the stability credentials of the adaptive estimation based formulation.

Theorem 2.1 *Under assumptions (A1), the control law (2.10), with adaptive laws (2.13), the inverse (2.6), and the plant (2.11), closed-loop signal boundedness is guaranteed and* $\lim_{t\to\infty} e(t) = 0$ *for* $e(t) = x(t) - x_r(t)$.

Proof Using (2.11), (2.12), and $\tilde{\theta}_a(t) = \theta_a(t) - \theta_a^*$, $\tilde{\theta}_b(t) = \theta_b(t) - \theta_b^*$, we have

$$\dot{e}(t) = (A - Bk)e(t) + B\left(\left(\frac{\theta_b(t) - u_d(t)}{\theta_a(t)}\right)\tilde{\theta}_a - \tilde{\theta}_b\right). \tag{2.16}$$

Apart from the tracking errors e, we have the parameter errors $\tilde{\theta}_a$ and $\tilde{\theta}_b$, and so we define a positive definite function

$$V(e, \tilde{\theta}_a, \tilde{\theta}_b) = e^T P e + \tilde{\theta}_a^2 \gamma_a^{-1} + \tilde{\theta}_b^2 \gamma_b^{-1}. \tag{2.17}$$

The time derivative of (2.17) is

$$\dot{V} = e^T(t) P \dot{e}(t) + \dot{e}^T(t) P e(t) + 2\tilde{\theta}_a(t)\gamma_a^{-1}\dot{\theta}_a(t) + 2\tilde{\theta}_b(t)\gamma_b^{-1}\dot{\theta}_b(t). \tag{2.18}$$

Substituting (2.13), (2.14) into (2.18), and noting that $\dot{\tilde{\theta}}_i(t) = \dot{\theta}_i(t)$, $i = a, b$, we get

$$\dot{V} = -e^T(t)Qe(t) + 2\tilde{\theta}_a(t)\gamma_a^{-1} f_a(t) + 2\tilde{\theta}_b(t)\gamma_b^{-1} f_b(t). \tag{2.19}$$

With $\theta_a(0) \geq \theta_a^a$, $\theta_b(0) \geq \theta_b^a$ and from the parameter projection algorithm (2.15), it follows that

$$\theta_i(t) \geq \theta_i^a, \quad \tilde{\theta}_i(t) f_i(t) \leq 0, i = a, b. \tag{2.20}$$

Next, we provide signal chasing analysis, to make appropriate conclusions regarding the stability of the system. With the properties (2.19) and (2.20), we get

$$\dot{V} = -e^T(t)Qe(t) \leq 0 \;\Rightarrow\; V(t) \in \mathcal{L}_\infty$$

$$\Downarrow$$

$$e(t) \in \mathcal{L}_\infty, \; \tilde{\theta}_a(t), \; \tilde{\theta}_b(t) \in \mathcal{L}_\infty$$

$$\Downarrow$$

$$x(t), \; \hat{\theta}_a(t), \; \hat{\theta}_b(t) \in \mathcal{L}_\infty$$

$$\Downarrow$$

$$u(t) \in \mathcal{L}_\infty \quad \text{(Adaptive Control Law)}$$

$$\Downarrow$$

$$\dot{e}(t) \in \mathcal{L}_\infty \quad \text{(Tracking Error Signal)}, \quad \dot{\hat{\theta}}_a(t), \dot{\hat{\theta}}_b(t) \in \mathcal{L}_\infty \text{(Adaptive Laws)}$$

$$\dot{V} = -e^T(t)Qe(t) \leq 0 \;\Rightarrow\; V(\infty) - V(0) \leq -\int_0^\infty e^T(t)Qe(t)dt \;\Rightarrow\; e(t) \in \mathcal{L}_2$$

Therefore, by Barbalet Lemma, $\lim_{t\to\infty} e(t) = 0$. ▽

New technical challenges unaddressed in the adaptive control literature [26, 27], like the parametrized function (2.5), the adaptive inverse (2.6), and the adaptive function $g_1(t)$ in (2.13) are addressed herewith. It is also clear that the adaptive inverse (2.6) presents saturation related issues, which were not encountered in previous adaptive inverse models presented in [26], [27].

For simulation-based analysis and study of the effectiveness of AIC, we make an assumption that $\theta_b = \theta_b^* = p_1$, i.e., $\tilde{\theta}_b = 0$, and so (2.8) reduces to

$$u_a(t) - u_d(t) = \frac{(\theta_b(t) - u_d(t))(\theta_a(t) - \theta_a^*)}{\theta_a(t)}, \tag{2.21}$$

and (2.11) reduces to

$$\dot{x}(t) = (A - Bk)x(t) + Br(t) + B\left(\frac{\theta_b(t) - u_d(t)}{\theta_a(t)}\right)\tilde{\theta}_a(t), \tag{2.22}$$

where $\tilde{\theta}_a(t) = \theta_a(t) - \theta_a^*$.

With $u_d(t) < \theta_b(t), \forall t \geq 0$, $\tilde{\theta}_b(t) = 0$ and the adaptive update signal for $\theta_a(t)$ (2.13), the plant (2.11) and corresponding reference (2.12), the proof in Theorem 2.1 is redone to ensure $\lim_{t\to\infty} e(t) = 0$.

2.5 Performance Evaluation of SJA

The simulation for the AIC scheme is performed with the states such as lateral velocity $x_1(t)$, roll rate $x_2(t)$, yaw rate $x_3(t)$ and roll angle $x_4(t)$. The matrices $A \in R^{4\times4}$ and $B \in R^{4\times1}$ [15, 23] are

$$A = \begin{bmatrix} -0.0134 & 48.5474 & -632.3724 & 32.0756 \\ -0.0199 & -0.1209 & 0.1628 & 0 \\ -0.0024 & -0.0526 & -0.0252 & 0 \\ 0 & 1 & 0.0768 & 0 \end{bmatrix}, \quad B = \begin{bmatrix} 0 \\ -0.0431 \\ -0.0076 \\ 0 \end{bmatrix},$$

which represents an unstable system. With a linear-quadratic regulator, $Q_1 = I_4$ and a constant $r_1 = 5$, we find an optimally controlled gain vector

$$k = \begin{bmatrix} 0.9070 & -106.9310 & 263.3615 & -16.3616 \end{bmatrix},$$

such that

$$A_r = \begin{bmatrix} -0.0134 & 48.5474 & -652.3724 & 32.0756 \\ 0.0192 & -4.7296 & 11.5137 & -0.7052 \\ 0.0045 & -0.8653 & 1.9763 & -0.1243 \\ 0 & 1 & 0.0768 & 0 \end{bmatrix},$$

which has closed loop poles $-1.1811, -0.7792 + 0.8684i, -0.7792 - 0.8684i$, and -0.0272, and high-performance design of the system. Using (2.14), we get

$$P = \begin{bmatrix} 0.755 & 48.004 & -314.065 & 16.838 \\ 48.004 & 27930.082 & -141015.824 & 7750.340 \\ -314.066 & -141015.824 & 722568.163 & -39605.211 \\ 16.838 & 7750.34 & -39605.211 & 2190.417 \end{bmatrix}.$$

For simulation, a bounded reference signal is chosen such that it changes significantly at 60 s. The objective is to ascertain robustness of the control action under a changed reference signal as well. We choose

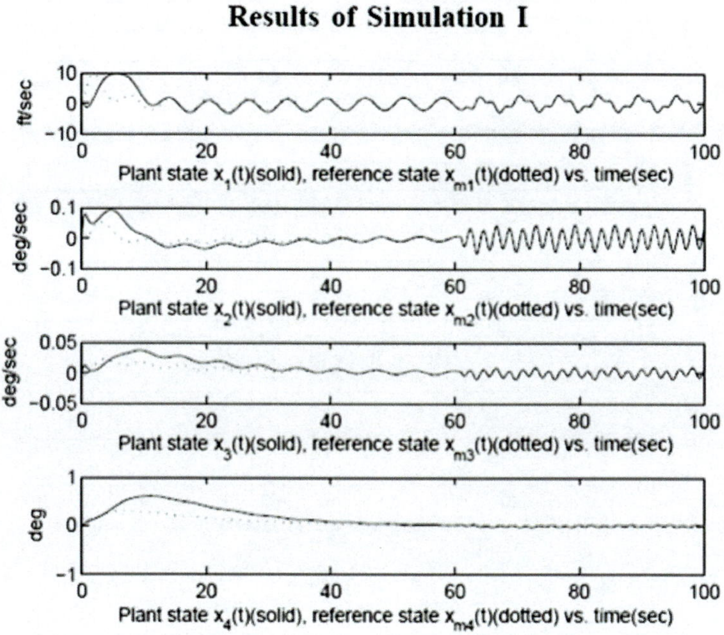

Fig. 2.8 Plant and reference states

$$r(t) = \begin{cases} 1.5\sin(t), & 0 \le t \le 60 \\ 1.5\sin(t) + 3\sin(3t), & t \ge 60. \end{cases}$$

Under assumptions (A1), $\theta_2 = 15°$ is assumed. The initial states are: lateral velocity (0.1 ft/s), roll rate (0.01°/s), yaw rate (0.001°/s) and roll angle (0.01°) for both the types of inverses. The initial reference states are: (0.5 ft/s), (0.1°/s), (0.01°/s) and (0.1°) respectively. An initial estimate $\theta_1(0) = 20$ v-° for the true nonlinearity parameter $\theta_a^* = 33.33$ v-°, lower bound of θ_a^* as $\theta_a^1 = 1$ v-° and adaptation gain $\gamma_a = 1$ are chosen.

Simulation I. The AIC results are provided in Figs. 2.8, 2.9 and 2.10. Specifically, Fig. 2.8 illustrates the state signals responses $x_i(t),\ i = 1, 2, 3, 4$ and the reference signals $x_{ri}(t),\ i = 1, 2, 3, 4$.

Figure 2.9 shows the asymptotic convergence of the tracking errors $e(t) = x_i(t) - x_{ri}(t)$. Figure 2.10 shows the input signal and the desired control law, and the asymptotic convergence of the parameter error $\theta_a(t) - \theta_a^*$ and the control error $u_a(t) - u_d(t)$.

The adaptive inverse approach does not guarantee parameter convergence and only guarantees asymptotic state tracking with parameter boundedness. In Simulation I, we apply a reference signal $r(t)$ under assumptions (A1) and (A2) for favorable AIC responses. However, when $r(t)$ is large, (A1) may be violated and the inverse function (2.6) may no longer be non-singular, and so we restrict $u_d(t)$ by re-defining a control law

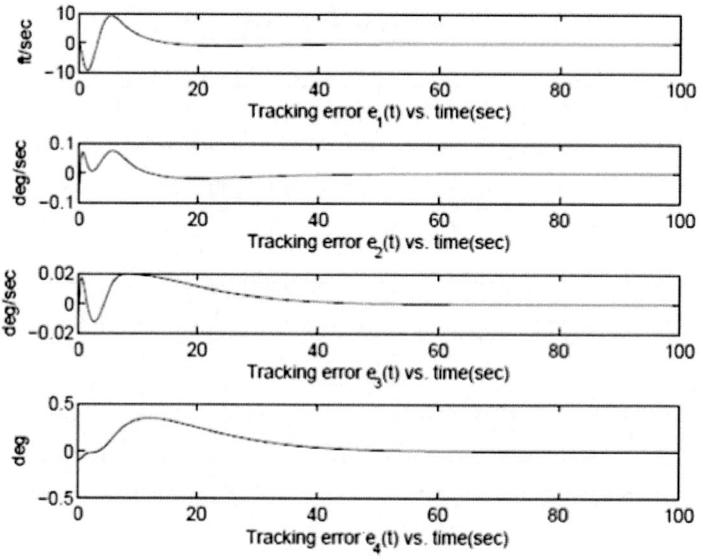

Fig. 2.9 State tracking errors

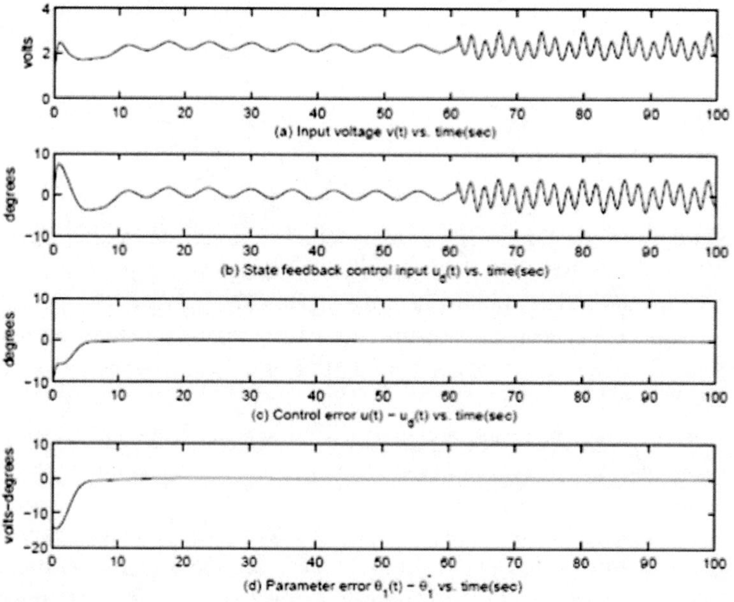

Fig. 2.10 **a** Input signal $v(t)$, **b** desired control signal $u_d(t)$, **c** control error $u_a(t) - u_d(t)$, **d** parameter error $\theta_a(t) - \theta_a^*$

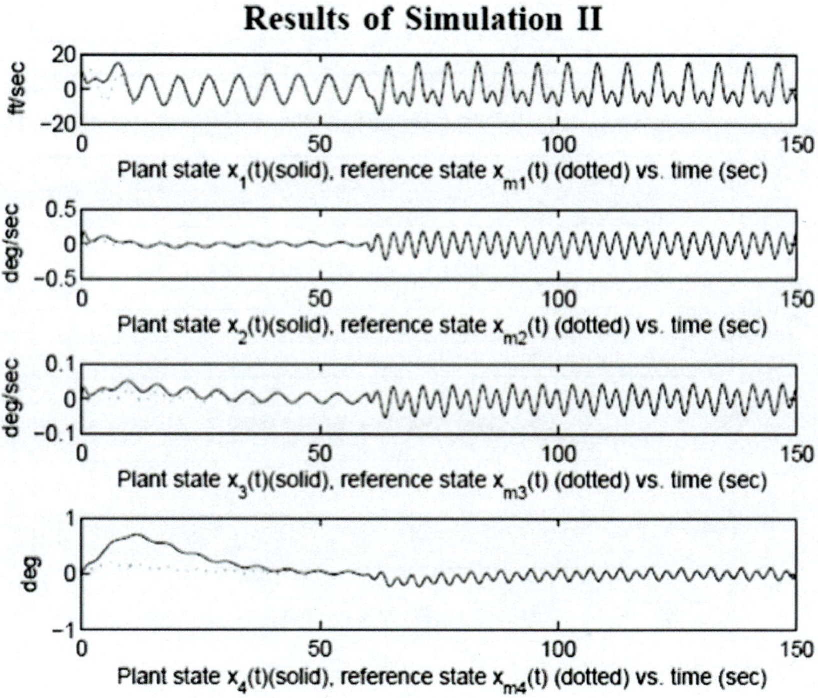

Fig. 2.11 Reference and plant states

$$\bar{u}_d(t) = -kx(t) + r(t), \tag{2.23}$$

with a small $\delta > 0$, such that

$$u_d(t) = \begin{cases} \theta_b - \delta & \text{if } \bar{u}_d(t) \geq \theta_b - \delta, \text{ and} \\ \bar{u}_d(t) & \text{otherwise.} \end{cases} \tag{2.24}$$

Simulation II.
For $\delta = 1$ and a reference signal $\bar{r}(t) = 5r(t)$, the simulation results are given in Figs. 2.11, 2.12 and 2.13. Specifically, Figs. 2.11 and 2.12 show the converge of plant states to the reference states, and that of the tracking errors to zero respectively. Figure 2.13 presents different signals clearly showing the affect of the change in the bounded reference signal at 60 s in all the signals such as the input voltage signal $v(t)$, the control signal $u_d(t)$, the adjusted control law under saturation $\bar{u}_d(t) - (\theta_b - \delta)$, and the parameter error signal $\tilde{\theta}_a(t)$.

Next, we present a possible distributed pattern and mathematical development of the SJAAs. We also propose and analyze the required AIC model arrays for the cancellation of the uncertainties arising from the SJAs. We also introduce a description of the challenges associated with the inversion of the actuator nonlinearities.

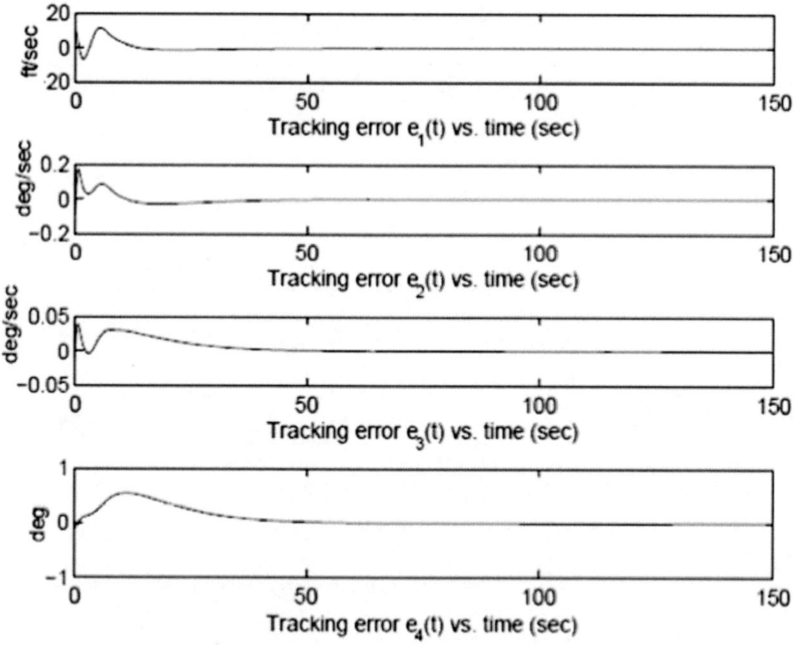

Fig. 2.12 State tracking errors (under saturation)

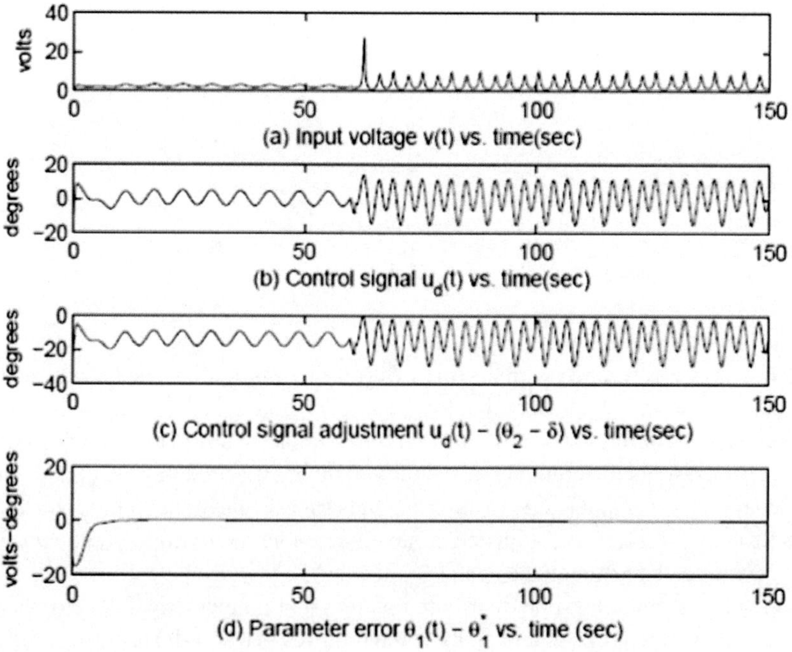

Fig. 2.13 **a** Input $v(t)$, **b** control law $u_a(t)$, **c** modified control $\bar{u}_d(t) - (\theta_b - \delta)$, and **d** parameter error $\theta_a - \theta_a^*$

2.6 Inverse Compensation for Actuator Arrays

An SJA is a miniature device and a single unit by itself does not generate enough flow realignment to cause any significant lift force for flight control. Additionally, all the surfaces like rudders, ailerons and elevators need to have SJAs distributed and implanted at specific locations to provide the necessary virtual shaping. Therefore, in the present study, we consider six location (three on either side of the longitudinal axis) for exploration in this direction.

Consider six arrays of SJAs with each jet characterized by a nonlinearity $N_{ij}(\cdot)$:

$$u_{aij}(t) = \bar{q} S\, C_{lij}(t) = N_{ij}(v_i(t)), \tag{2.25}$$

where u_{aij} is the effective lift force or the equivalent virtual deflection due to jth actuator in the ith array, S is the relevant surface area, $\bar{q}$ is the fluid dynamic pressure, $v_i(t) = A_{ipp}^2$ with $A_{ipp}(t)$ being the input peak-to-peak voltage amplitude applied to all the actuators in the array.

The parametric model for a SJA [15, 16] and the actuation of the SJs, generate a control law to suitably modify the airfoil virtual shape. With flow control at constant airstream density, an empirical model developed through experiments on the SJs is

$$C_{lij} = p_{1ij} - \frac{p_{2ij} p_{3ij} u_{\infty ij}}{f c C_{\mu ij}(t)}, \quad i = 1, \ldots, 6, \;\; j = 1, \ldots, n_i, \tag{2.26}$$

where n_i represents the number of actuators in the ith array, $C_{\mu ij}(t) = \frac{p_{4ij} A_{ppi}^2(t)}{u_{\infty ij}^2}$, c is the local aerodynamic chord, f is the frequency, the parameters p_{kij}, $k = 1, \ldots, 4, i = 1, \ldots, 6,\; j = 1, \ldots, n_i$, are the uncertain positive scalar values and are obtained from a baseline model based on the virtual shaping effect on a real 2-D wing.

By use of (2.26), the airfoil surface deflection due to the SJAs on the airfoil surface is given by

$$u_{aij} = \bar{q} S p_{1ij} - \bar{q} S \frac{p_{2ij} p_{3ij} u_{\infty ij}}{f c C_{\mu ij}(t)}, \quad i = 1, \ldots, 6, \;\; j = 1, \ldots, n_i. \tag{2.27}$$

In order to perform adaptive compensation, a parametric model is obtained with parameters represented as $\theta_{ij}^* = [\theta_{1ij}^*, \theta_{2ij}^*]^T$. The parametrized model for the jth SJA in the ith array is

$$u_{aij}(t) = \theta_{bij}^* - \frac{\theta_{aij}^*}{v_i(t)}, i = 1, \ldots, 6, \;\; j = 1, \ldots, n_i, \tag{2.28}$$

where $\theta_{bij}^* = p_{1ij}$, $\theta_{aij}^* = \frac{p_{2ij} p_{3ij} u_{\infty ij}^3}{f c p_{4ij} A_{ipp}^2(t)}$. A reformulation is given as

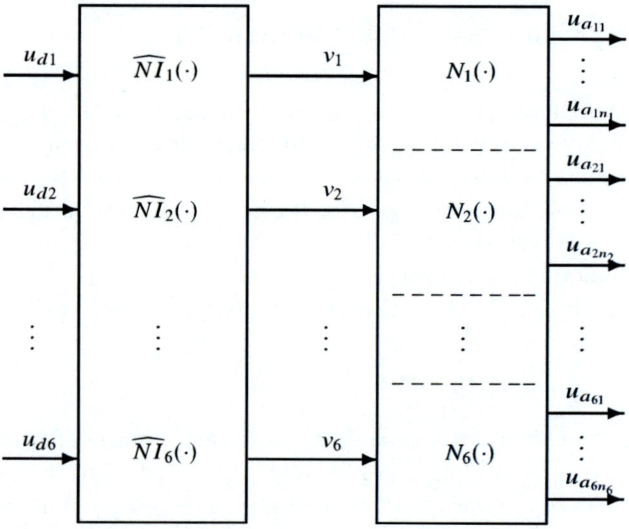

Fig. 2.14 Adaptive compensation for SJ arrays

$$u_{ai}(t) = \sum_{j=1}^{n_i} \alpha^*_{ij} u_{aij}(t) = \theta^*_{bi} - \frac{\theta^*_{ai}}{v_i(t)}, i = 1, 2, \ldots, 6, \tag{2.29}$$

where α^*_{ij} are constants, $\theta^*_i = [\theta^*_{ai}, \theta^*_{bi}]^T$, $\theta^*_{bi} = \sum_{j=1}^{n_i} \alpha^*_{ij}\theta^*_{bij}$, $\theta^*_{ai} = \sum_{j=1}^{n_i} \alpha^*_{ij}\theta^*_{aij}$, and $u_{ai}(t)$ represents the cumulative virtual deflection from the ith array.

As explained in the case of the single SJA, the uncertainties need to be compensated using an adaptive inverse function. Next, we undertake the same process for these distributed SJAAs. The essence of the AIC approach is to employ an inverse

$$v_i(t) = \widehat{NI}_i(u_{di}(t)) \tag{2.30}$$

to cancel the effect of the unknown nonlinearity $N_i(\cdot)$, where the inverse characteristic $\widehat{NI}_i(\cdot)$ is parametrized by an estimate $\theta_i(t) = [\theta_{ai}(t), \theta_{bi}(t)]^T$ of θ^*_i and $u_{di}(t)$ is a desired control law. The appropriate adaptive inverse for (2.29) is given by

$$v_i(t) = \frac{\theta_{ai}(t)}{\theta_{bi}(t) - u_{di}(t)}, \quad i = 1, \ldots, 6, \tag{2.31}$$

where $\theta_{ai}(t), \theta_{bi}(t)$ are to be estimated from adaptive update such that $\theta_{ai}(t), \theta_{bi}(t) > 0$, because $\theta^*_{ai}, \theta^*_{bi} > 0$, needed in order to implement the inverses, assuming $u_{di}(t) < \theta_{bi}(t)$, $\forall t \geq 0$, $i = 1, \ldots, 6$ where $\theta_{bi}(t)$ represents the maximum deflection.

An adaptive compensation framework with the arrays of SJ nonlinearities and the inverse jet arrays are represented as follows in Fig. 2.14.

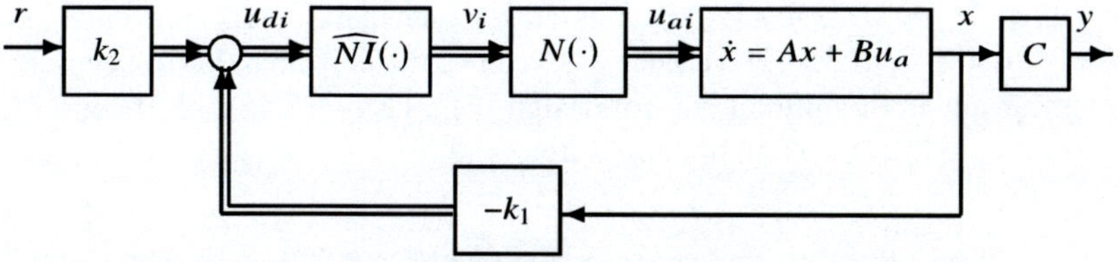

Fig. 2.15 State feedback inverse control system

2.6.1 Control Error Formulation

From (2.29) and (2.31), we get

$$u_{ai}(t) = \frac{\theta^*_{bi}\theta_{ai}(t) - \theta^*_{ai}(\theta_{bi}(t) - u_{di}(t))}{\theta_{ai}(t)}. \tag{2.32}$$

Thus, the control error is expressed as

$$u_{ai}(t) - u_{di}(t) = \frac{\theta^*_{bi}\theta_{ai}(t) - \theta^*_{ai}\theta_{bi}(t) + u_{di}(t)\theta^*_{ai} - u_{di}(t)\theta_{1i}(t)}{\theta_{1i}(t)}.$$

Rearranging, we get

$$u_{ai}(t) - u_{di}(t) = \left(\frac{\theta_{bi}(t) - v_{bi}(t)}{\theta_{ai}(t)}\right)\tilde{\theta}_{ai}(t) - \tilde{\theta}_{bi}(t), \tag{2.33}$$

where $\tilde{\theta}_{ai} = \theta_{ai}(t) - \theta^*_{ai}$, and $\tilde{\theta}_{bi} = \theta_{bi}(t) - \theta^*_{bi}$.

Next, we present an AIC in state feedback mode for cancellation of $N_i(\theta^*_i; v_i(t))$, $\forall i = 1, \ldots, 6$, in order to meet the control objective. Such a control system is shown in Fig. 2.15.

With a controllable and linear aircraft model, we have

$$\dot{x}(t) = Ax(t) + Bu_a(t) = Ax(t) + \sum_{i=1}^{6}\sum_{j=1}^{n_i} b_{ij}u_{aij} = Ax(t) + \sum_{i=1}^{6} b_i u_{ai}, \; x \in R^n, \tag{2.34}$$

where $B = [b_1, \ldots, b_6] \in R^{n\times 6}$, provided we assume that

$$b_{ij} = \xi_{ij}b_{ik}, \; j \neq k, \; j, k \in 1, 2, \ldots, n_i,$$

where ξ_{ij} are proportionality constants. For system stability, the desired control law is given by

$$u_d(t) = -k_1^T x(t) + k_2 r(t), \tag{2.35}$$

where $u_d(t) = [u_{d1}, \ldots, u_{d6}]^T$, and k_1 is chosen such that $A - Bk_1$ are equal to some desired closed-loop poles. The choice of k_1 can be made from a Pole placement [24] or a Linear quadratic optimal control design [21]. Using (2.46), (2.35) and (2.31), the closed-loop linear system dynamics are

$$\dot{x}(t) = (A - Bk_1)x(t) + Bk_2 r(t) + \sum_{i=1}^{6} b_i \left(\frac{\theta_{bi}(t) - u_{di}(t)}{\theta_{ai}(t)} \tilde{\theta}_{ai}(t) - \tilde{\theta}_{bi}(t) \right), \quad (2.36)$$

where $\sum_{i=1}^{6} b_i k_{1i}^T = Bk_1$, $\sum_{i=1}^{6} b_i k_{2i} = Bk_2$.

While maintaining a similar structure with the known components, consider a reference system given as

$$\dot{x}_m(t) = (A - Bk_1)x_r(t) + Bk_2 r(t), \quad (2.37)$$

where $x_r(t)$ represents the desired aircraft state trajectory and the body axis dynamics are stabilized by the stable matrix $A - Bk_1$. The physical meaning of $Bk_2 r(t)$ is the desired external body axis motion command.

2.6.2 *Adaptive Estimation of Uncertain Parameters*

Assuming some known lower bounds for all the uncertain parameters, the characteristic $N(\theta_i(t); v_i(t))$ provides a similar nonlinearity model to $N(\theta_i^*; v_i(t))$ given in (2.29). Initial estimates $\theta_{1i}(0)$ of θ_{1i}^* and $\theta_{2i}(0)$ of θ_{2i}^* are chosen such that $\theta_{1i}(0) \geq \theta_{1i}^a$ and $\theta_{2i}(0) \geq \theta_{2i}^a$ respectively. We formulate the update laws for θ_{li} as

$$\dot{\theta}_{li}(t) = g_{li}(t) + f_{li}(t), \; l = 1, 2, \; i = 1, \ldots, 6, \; t \geq 0, \quad (2.38)$$

where $f_{li}(t)$ are projection functions to be determined, and the functions $g_{li}(t)$ are

$$g_{1i}(t) = -\gamma_{1i} e^T(t) P b_i \left(\frac{\theta_{2i}(t) - v_{2i}(t)}{\theta_{1i}} \right), \quad g_{2i}(t) = \gamma_{2i} e^T(t) P b_i,$$

with some design parameters $\gamma_{1i}, \gamma_{2i} > 0$ to control adaptation rates, and $P \in R^{n \times n}$ which is a constant matrix such that $P = P^T > 0$ satisfies the Lyapunov equation

$$P(A - Bk_1) + (A - Bk_1)^T P = -Q. \quad (2.39)$$

For some constant $Q = Q^T > 0$, with any given Q, such a P can be determined uniquely. The purpose and merit of using projection functions has already been described. For adaptive laws (2.38), the projection functions $f_{li}(t)$ are

$$f_{li}(t) = \begin{cases} 0 & \text{if} \quad \theta_{li}(t) > \theta_{li}^a, \text{ or} \\ & \text{if} \quad \theta_{li}(t) = \theta_{li}^a \text{ and } g_{li}(t) \geq 0, l = 1, 2, \ i = 1, \ldots, 6, \\ -g_{li}(t) & \text{otherwise.} \end{cases} \tag{2.40}$$

The implementation of the adaptive laws for $\theta_{li}(t)$ are simple:

$$\theta_{li}(t) = -\int_0^t (g_{li}(\tau) + f_{li}(\tau))\, d\tau + \theta_{li}(0).$$

Theorem 2.2 *Under assumption (A1), the control laws (2.35), updated from the adaptive laws (2.38), the inverses (2.31) and applied to the plant (2.46), guarantee closed-loop signal boundedness and* $\lim_{t\to\infty} e(t) = 0$.

Proof Using (2.36) and (2.37), the dynamic equation for tracking error $e(t)$ is

$$\dot{e}(t) = (A - Bk_1)e(t) + \sum_{i=1}^{6} b_i \left(\frac{\theta_{2i}(t) - v_{2i}(t)}{\theta_{1i}(t)} \tilde{\theta}_{1i}(t) - \tilde{\theta}_{2i}(t) \right). \tag{2.41}$$

Let us consider a function which is scalar continuous and positive definite, and expressed as

$$V = e^T Pe + \sum_{j=1}^{2} \sum_{i=1}^{6} \tilde{\theta}_{ji}^2 \gamma_{ji}^{-1}, \tag{2.42}$$

as a measure of $e, \tilde{\theta}_{li}, \ t \geq 0$. Noting that $\dot{\tilde{\theta}}_{li}(t) = \dot{\theta}_{li}(t)$. The time derivative of V is given by

$$\dot{V} = e^T(t)P\dot{e}(t) + \dot{e}^T(t)Pe(t) + 2\sum_{l=1}^{2} \sum_{i=1}^{6} \tilde{\theta}_{li}(t)\gamma_{li}^{-1}\dot{\theta}_{li}(t). \tag{2.43}$$

Substituting (2.38), (2.39) and (2.41) into (2.43), we have

$$\dot{V} = -e^T(t)Qe(t) + 2\sum_{l=1}^{2} \sum_{i=1}^{6} \tilde{\theta}_{li}(t)\gamma_{li}^{-1} f_{li}(t).$$

With $\theta_{li}(0) \geq \theta_{li}^a$, and the projection function (2.40), we get

$$\theta_{li}(t) \geq \theta_{li}^a, \ \ \tilde{\theta}_{li}(t) f_{li}(t) \leq 0, \ \ l = 1, 2, i = 1, \ldots, 6. \tag{2.44}$$

With this property, (2.6.2) reduces to the negative semi-definite time-derivative

$$\dot{V} = -e^T(t)Qe(t) \leq 0 \ \Rightarrow \ V(t) \in \mathcal{L}_\infty \tag{2.45}$$

$$\Downarrow$$

$$e(t) \in \mathcal{L}_\infty,\ \tilde{\theta}_{1i}(t),\ \tilde{\theta}_{2i}(t) \in \mathcal{L}_\infty$$

$$\Downarrow$$

$$x(t),\ \hat{\theta}_{1i}(t),\ \hat{\theta}_{2i}(t) \in \mathcal{L}_\infty$$

$$\Downarrow$$

$$u(t) \in \mathcal{L}_\infty \quad \text{(Adaptive Control Law (2.35))}$$

$$\Downarrow$$

$$\dot{e}(t) \in \mathcal{L}_\infty \quad \text{(Tracking Error Signal (2.41))},$$

$$\dot{\hat{\theta}}_{1i}(t), \dot{\hat{\theta}}_{2i}(t) \in \mathcal{L}_\infty \text{(Adaptive Laws (2.38))}$$

$$\dot{V} = -e^T(t)Qe(t) \leq 0 \ \Rightarrow \ V(\infty) - V(0) \leq -\int_0^\infty e^T(t)Qe(t)dt \ \Rightarrow \ e(t) \in \mathcal{L}_2$$

Therefore, by Barbalet Lemma, $\lim_{t\to\infty} e(t) = 0$. ∇

2.6.3 *Robustness Issues*

One can expect a change in the actuator performance with some gain uncertainty due to an unexpected cross-flow impeding the normal flow re-circulation region created by SJA control; in sharp contrast to when the actuator is calibrated in a quiescent environment. In Fig. 2.16, we represent SJAs acting on an aircraft model in the presence of gain uncertainties and disturbances. The gain of the disturbance is being adaptively estimated.

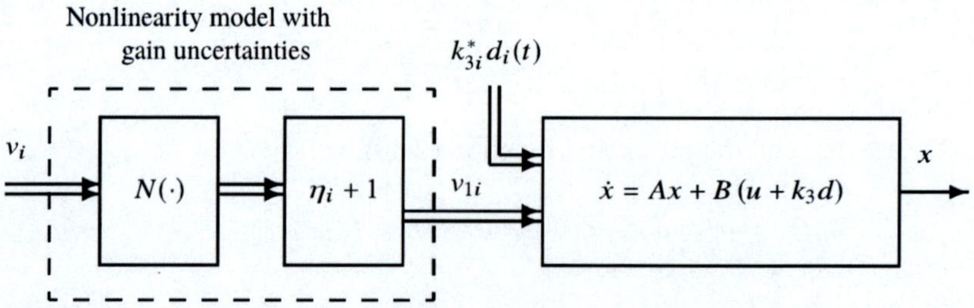

Fig. 2.16 Aircraft model with jet nonlinearities in presence of gain uncertainties and disturbances

We consider aircraft dynamics with modeling errors described by

$$\dot{x}(t) = Ax(t) + \sum_{i=1}^{6} b_i u_i(t) + \sum_{i=1}^{6} b_i k_{3i}^* d_i(t), \; x \in R^n, \tag{2.46}$$

where $[b_1, \ldots, b_6] = B$. The desired state feedback control signal is modified as

$$u_d(t) = -k_1^T x(t) + k_2 r(t) - k_3(t)d(t), \tag{2.47}$$

where $r(t)$ is bounded, $k_3(t)d(t) = [k_{31}(t)d_1(t), \ldots, k_{36}(t)d_6(t)]^T$, are bounded. Using (2.33) and (2.46), the closed-loop linear system dynamics are

$$\begin{aligned}\dot{x}(t) &= Ax(t) + \sum_{i=1}^{6} b_i \left(v_{2i}(t) + \frac{\theta_{2i}(t) - v_{2i}(t)}{\theta_{1i}(t)} \tilde{\theta}_{1i}(t) - \tilde{\theta}_{2i}(t) \right) + \sum_{i=1}^{6} b_i k_{3i}^* d_i(t) \\ &= (A - Bk_1)x(t) + Bk_2 r(t) + \sum_{i=1}^{6} b_i \left(\frac{\theta_{2i}(t) - v_{2i}(t)}{\theta_{1i}(t)} \tilde{\theta}_{1i} - \tilde{\theta}_{2i} \right) - B\tilde{k}_3 d(t), \end{aligned} \tag{2.48}$$

where $\sum_{i=1}^{6} b_i k_{1i}^T = Bk_1$, $\sum_{i=1}^{6} b_i k_{2i} = Bk_2$, $\sum_{i=1}^{6} b_i \tilde{k}_{3i} d_i(t) = B k_3 d(t)$, $\tilde{k}_{3i}(t) = k_{3i}(t) - k_{3i}^*$. For a reference system (2.37), the objective is to formulate the matrix k_1, the vector k_2 and the adaptive laws for $\theta_{1i}(t)$, $\theta_{2i}(t)$, $k_{3i}(t)$, such that closed-loop signals are bounded and $\lim_{t\to\infty}(x(t) - x_r(t)) = 0$.

We formulate an additional adaptive update law for $k_{3i}(t)$ as

$$\dot{k}_{3i}(t) = \gamma_{3i} e^T(t) P b_i d_i(t), \; i = 1, \ldots, 6, \tag{2.49}$$

where $\gamma_{3i} > 0$ and $P = P^T > 0$ satisfies (2.39).

Theorem 2.3 *Under assumption (A1), the control laws (2.47), updated from the adaptive laws (2.38), (2.49) the inverses (2.31) applied to the plant (2.46), guarantee closed-loop signal boundedness and* $\lim_{t\to\infty} e(t) = 0$, *in presence of gain uncertainties and bounded disturbances as shown in Fig. 2.15.*

Proof Using (2.48) and (2.37), we have

$$\dot{e}(t) = (A - Bk_1)e(t) + \sum_{i=1}^{6} b_i \left(\frac{\theta_{2i}(t) - v_{2i}(t)}{\theta_{1i}(t)} \tilde{\theta}_{1i} - \tilde{\theta}_{2i} \right) - B\tilde{k}_3 d(t). \tag{2.50}$$

Consider the positive definite function

$$V = e^T P e + \sum_{j=1}^{2} \sum_{i=1}^{6} \tilde{\theta}_{ji}^2 \gamma_{ji}^{-1} + \sum_{i=1}^{6} \tilde{k}_{3i}^2 \gamma_{3i}^{-1}, \tag{2.51}$$

as a measure of e, $\tilde{\theta}_{li}$. Noting that $\dot{\tilde{\theta}}_{li}(t)=\dot{\theta}_{li}(t)$, $\dot{\tilde{k}}_{3i}(t) = \dot{k}_{3i}(t)$, $l = 1, 2$, $i=1, \ldots, 6$. The time derivative of V is given by

$$\dot{V} = e^T(t)P\dot{e}(t) + \dot{e}^T(t)Pe(t) + 2\sum_{l=1}^{2}\sum_{i=1}^{6}\tilde{\theta}_{li}(t)\gamma_{li}^{-1}\dot{\theta}_{li}(t) + 2\sum_{i=1}^{6}\tilde{k}_{3i}(t)\gamma_{3i}^{-1}\dot{k}_{3i}(t). \tag{2.52}$$

Substituting (2.39), (2.38) and (2.50) into (2.52), we have

$$\dot{V} = -e^T(t)Qe(t) + 2e^T(t)P\sum_{i=1}^{6}b_i\left(\frac{\theta_{2i}(t) - v_{2i}(t)}{\theta_{1i}(t)}\tilde{\theta}_{1i} - \tilde{\theta}_{2i}\right) + 2\sum_{l=1}^{2}\sum_{i=1}^{6}\tilde{\theta}_{li}(t)\gamma_{li}^{-1}\dot{\theta}_{li}(t). \tag{2.53}$$

Therefore, substituting (2.38) in (2.53), we get (2.6.2), and noting that $k_{3i}(t)$ are bounded, from (2.50) we get that $\dot{e}(t)$ is bounded, and following the remaining steps of signal chasing, we have $\lim_{t\to\infty} e(t) = 0$, in presence of gain uncertainties and bounded disturbances as shown in Fig. 2.15. ∇

We find analytically that the AIC framework effectively compensates the uncertainties in SJAs. Next, we provide performance validation through a simulation-based study of a 3 DOF (degrees of freedom) linearized model.

2.7 System Performance Evaluation with SJAA

A likely platform for the possible application of the virtual shape change technology is the next-generation stealth aircraft like the Barron Associates' medium-fidelity Uninhabited Combat Air Vehicle (UCAV) model by utilizing only the available non-proprietary details. The BANTAM is a blended-wing-body design which adequately encapsulates the described actuator uncertainties. The physical parameters of the BANTAM are summarized in Table 2.1.

Table 2.1 Physical attributes of BANTAM

Parameter	Value
Leading edge sweep	50°
Mass	419.3 Slug
Weight	13,500 lb
Span	32 ft
Reference area	410 ft^2
Aspect ratio	2.5
$[I_{xx}\ \ I_{yy}\ \ I_{zz}]$	[11000 6500 18500] slug-ft^2
Mean aerodynamic chord	15.3 ft

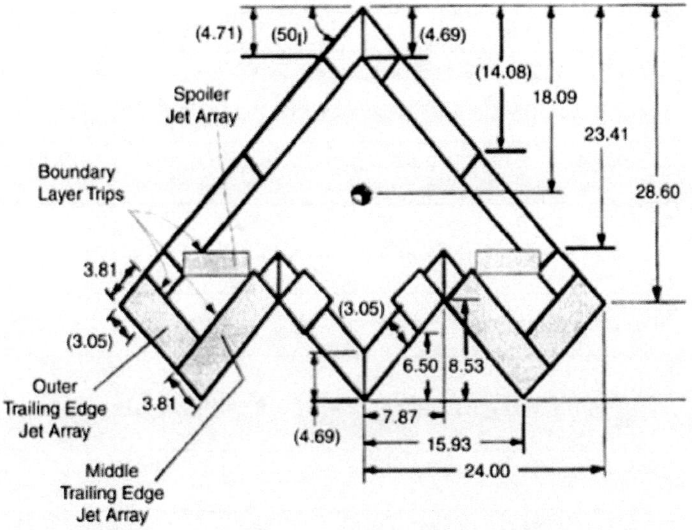

Fig. 2.17 BANTAM airfoil with six SJAAs (measurements in feet)

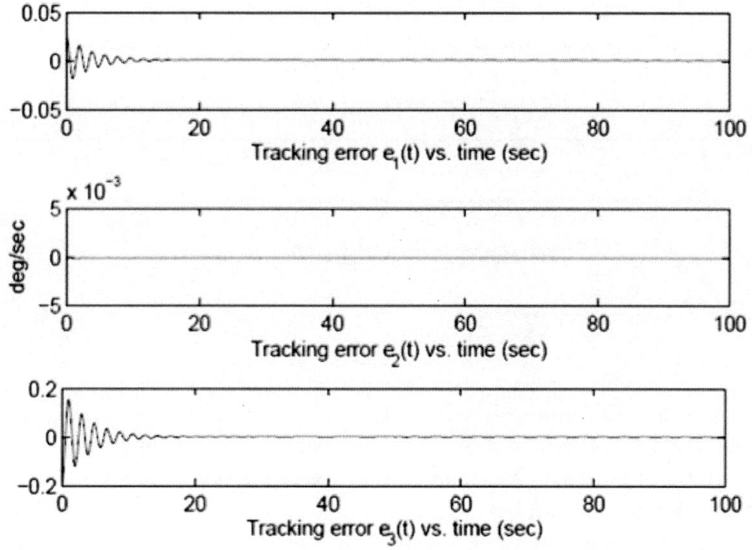

Fig. 2.18 Tracking errors $e_i(t)$

For a system performance evaluation of this configuration, the six SJAA (six shaded regions) can replace the model trailing edge flaps and the spoilers. The distribution is symmetric on both sides of the longitudinal axis as shown in Fig. 2.17. The specific location of SJAs is vital for generation of appropriate distributed and localized airfoil surface pressure distribution.

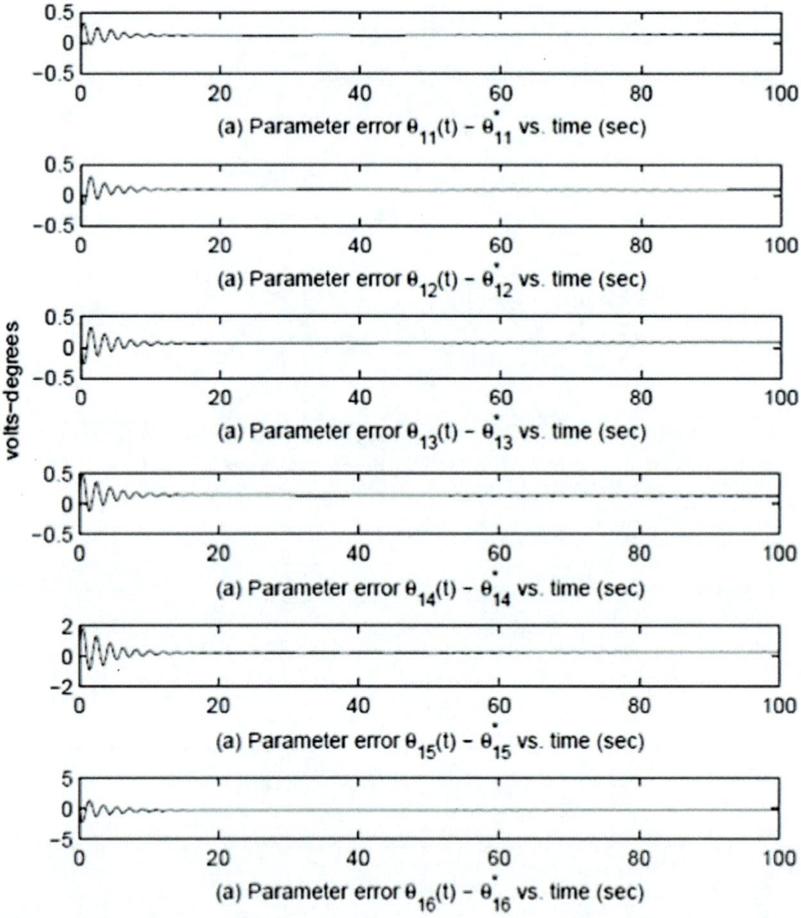

Fig. 2.19 Parameter errors $\tilde{\theta}_{1i}(t)$

The AIC technique applies to a 3-DOF model with six arrays of SJAs- three on either side of the longitudinal axis. We demonstrate the feasibility of nonlinearity compensation of the SJAAs by using a linear quadratic regulator (LQR) design for controller evaluation. We undertake a simulation-based analysis for the AIC scheme on a 3-DOF linearized aircraft (BANTAM) model for the states, $x = [x_1, x_2, x_3]^T$ at the trim condition: $M = 0.455, \alpha = 2.7°$, and $\beta = 0$. The matrices $A \in R^{3\times3}$ and $B \in R^{3\times6}$ are obtained analytically for the BANTAM with mechanical control surfaces. The matrices A and B, the appropriate initial conditions, and appropriate reference signal are all provided along with the two true nonlinearity parameters for all the six arrays along with the corresponding initial estimates [3]. The algebraic riccati equation is solved for an unique stabilizing solution matrix leading to the gain matrix for stable closed-loop poles. The CARE function in Matlab ensures that the poles of A that are stable are almost same as those of $A - BK_1$. The appropriate

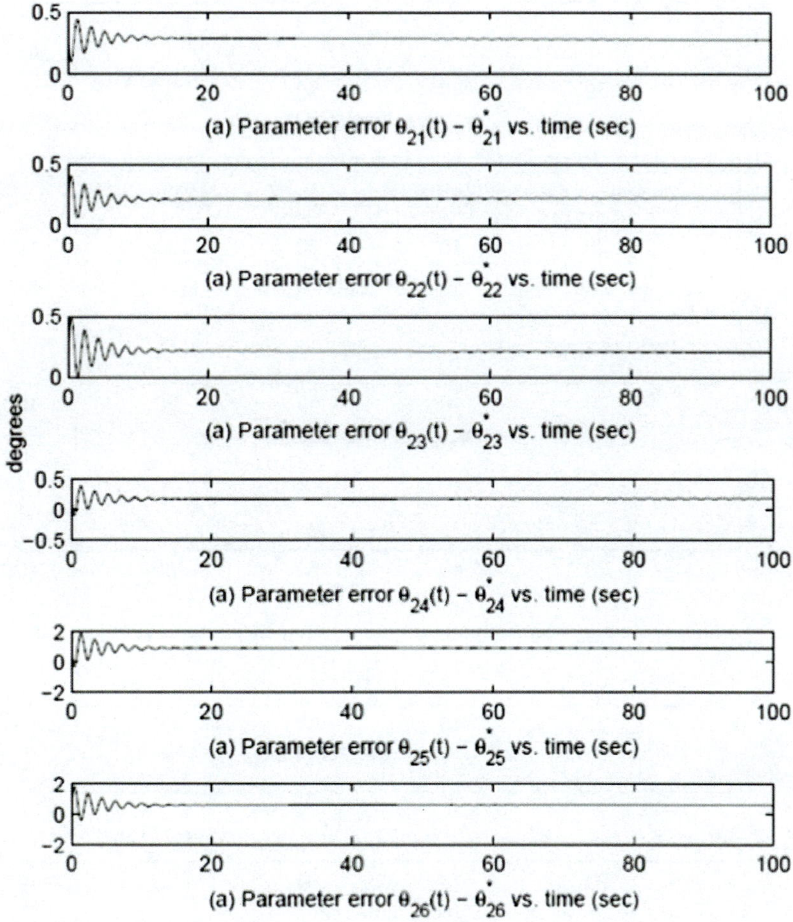

Fig. 2.20 Parameter errors $\tilde{\theta}_{2i}(t)$

adaptive gains are chosen, and the reference signal is purposely changed at 50 s into the simulation to impact the input signals as shown in Fig. 2.21, for instance.

In the simulation studies, we illustrate that with time-varying and bounded reference inputs that change at arbitrary instants of time, but don't violate the assumption (A1), asymptotic tracking is possible. We present simulation results for the asymptotic tracking of the reference states $x_m(t)$ by the system states $x(t)$. The results of adaptive inverse compensation are shown in Figs. 2.18, 2.19, 2.20, 2.21 and 2.22. Specifically, Fig. 2.18 shows that $e(t) = x(t) - x_m(t)$ approach zero asymptotically following brief transience, thereby establishing the robustness of the formulation presented.

Figures 2.19 and 2.20 show the convergence characteristics of the two types of parameter errors for the SJAAs to fixed values (parameter boundedness) which is

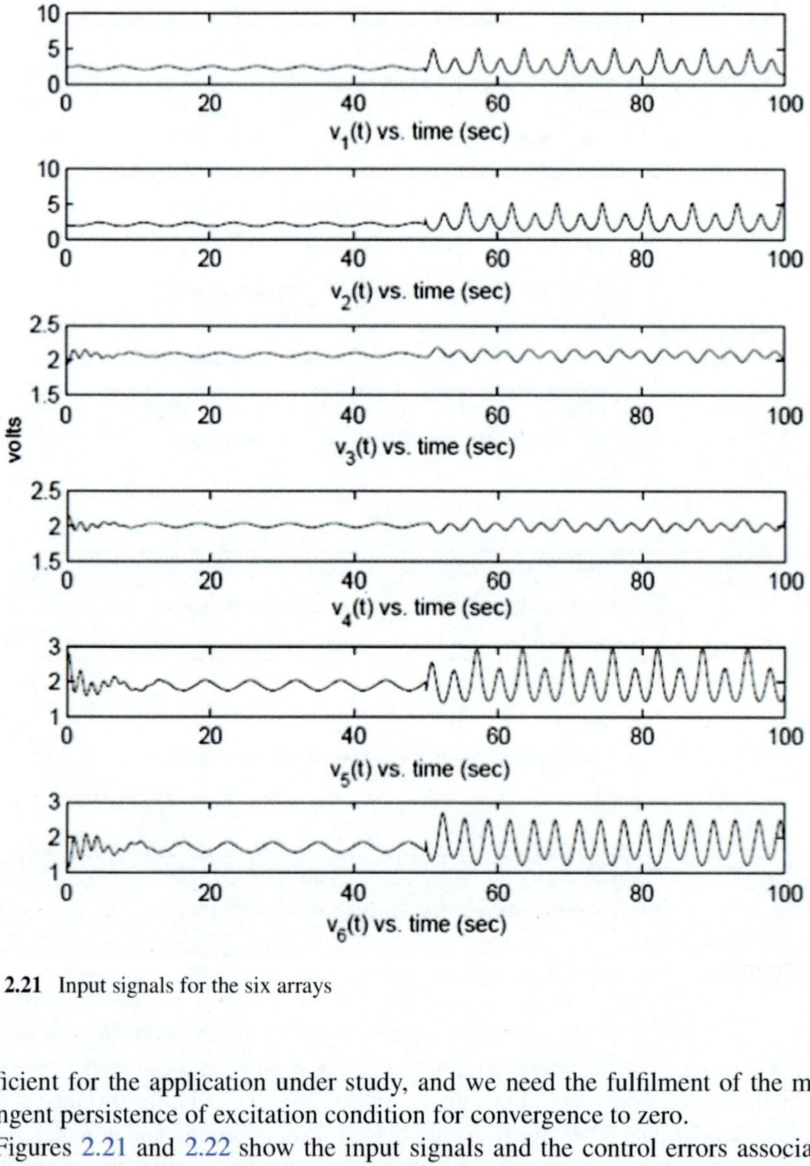

Fig. 2.21 Input signals for the six arrays

sufficient for the application under study, and we need the fulfilment of the more stringent persistence of excitation condition for convergence to zero.

Figures 2.21 and 2.22 show the input signals and the control errors associated with SJAAs respectively. The dependence of the signals $v_i(t)$ and $u_i(t) - u_{di}(t)$ on $r(t)$ is in the form of sinusoidal behavior. The change in such characteristics after 50 s is also attributable to $r(t)$, and the boundedness of these signals contribute to the asymptotic tracking of the state errors. Unlike the tracking errors, the parameter errors and the control error signals do not converge to zero, and such an arrangement is adequate for the present application.

In this chapter, we formulate an AIC scheme for flow rearrangement and separation control at low AoA that results in virtual shaping with SJAAs. A mathematical model of six SJAAs distributed symmetrically about the aircraft longitudinal axis

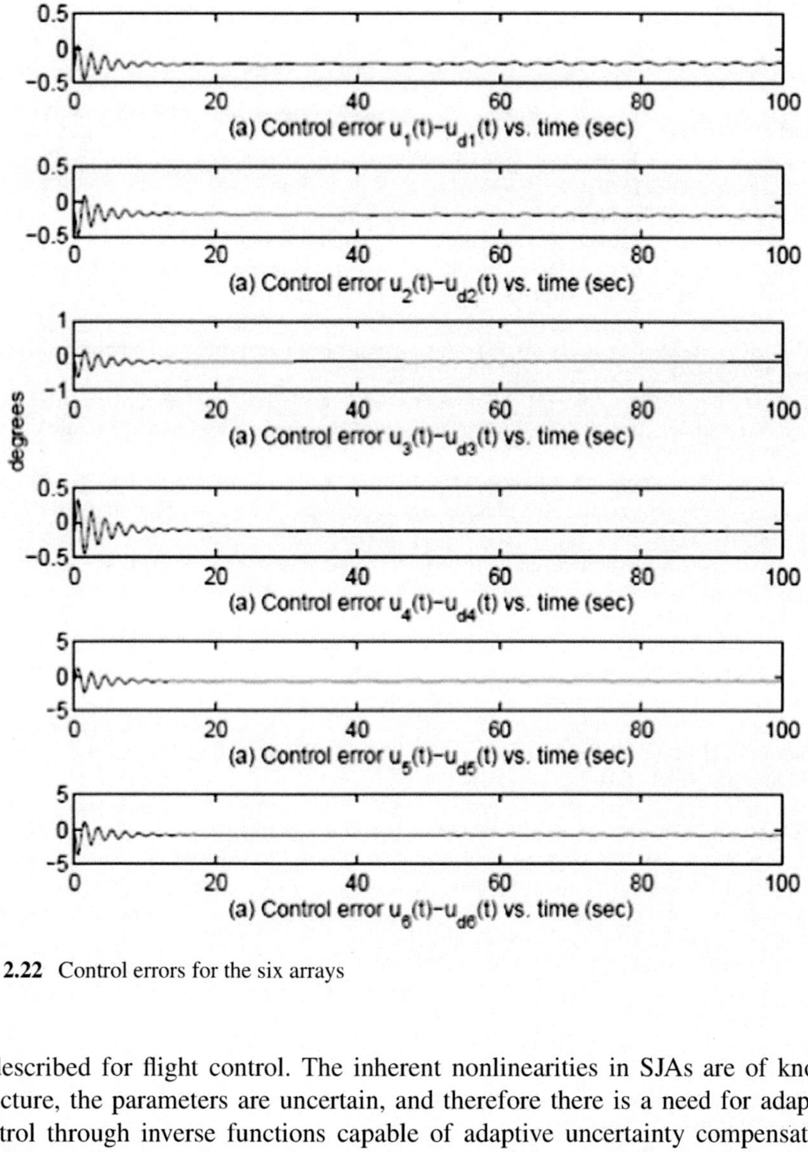

Fig. 2.22 Control errors for the six arrays

is described for flight control. The inherent nonlinearities in SJAs are of known structure, the parameters are uncertain, and therefore there is a need for adaptive control through inverse functions capable of adaptive uncertainty compensation. Subsequently, we formulate robust control laws and estimation based update laws to analyze stability and tracking performance under gain uncertainties and bounded disturbances. We find promising simulation results for tracking of the linearized tailless aircraft model.

References

1. Pack, L., Seifert, A.: Dynamics of active separation control at high Reynolds numbers. AIAA Paper No. 2000-0409, 38th Aerospace Sciences Meeting & Exhibit, Reno, NV (2000)
2. Deb, D., Tao, G., Burkholder, J., Smith, D.: An adaptive inverse control scheme for a synthetic jet actuator model. In: Proceedings of the 2005 American Control Conference. https://doi.org/10.1109/acc.2005.1470367
3. Deb, D., Tao, G., Burkholder, J., Smith, D.: An adaptive inverse control scheme for synthetic jet actuator arrays. Infotech@Aerospace (2005). https://doi.org/10.2514/6.2005-7170
4. Deb, D., Tao, G., Burkholder, J., Smith, D.: Adaptive compensation control of synthetic jet actuator arrays for airfoil virtual shaping. J. Aircr. **44**(2), 616–626 (2007). https://doi.org/10.2514/1.24910
5. Patel, R., Deb, D., Dey, R., Balas, E.V.: Adaptive and intelligent control of microbial fuel cells. Intell. Syst. Ref. Lib. (2020). https://doi.org/10.1007/978-3-030-18068-3
6. Deb, D., Tao, G., Burkholder, J.: High-order design of adaptive inverses for signal-dependent actuator nonlinearities. In: 2008 American Control Conference (2008). https://doi.org/10.1109/acc.2008.4587029
7. Chatlynne, E., Rumigny, N., Amitay, M., Glezer, A.: Virtual aero-shaping of a Clark-Y airfoil using synthetic jet actuators. In: 39th Aerospace Sciences Meeting and Exhibit (2001). https://doi.org/10.2514/6.2001-732
8. Zhang, J., Tan, X.: Experimental study on flow and heat transfer characteristics of synthetic jet driven by piezoelectric actuator. Sci China Ser E **50**, 221–229 (2007). https://doi.org/10.1007/s11431-005-0006-1
9. Gilmore, P., Sundaresan, V., Seidt, J., Smith, J.: Design and analysis of a synthetic jet actuator-based fluid atomization device. J. Intell. Mater. Syst. Struct. **28**(17), 2307–2316. SAGE Publications (2017). https://doi.org/10.1177/1045389x17689938
10. Feero, M. A., Lavoie, P., Sullivan, P.E.: Influence of cavity shape on synthetic jet performance. Sensors and Actuators A: Physical, vol. 223. Elsevier BV, pp. 1–10 (2015). https://doi.org/10.1016/j.sna.2014.12.004
11. Kim, S.H., Kim, C.: Separation control on NACA23012 using synthetic jet. Aerosp. Sci. Technol. **13**(4-5), 172–182. Elsevier BV (2009). https://doi.org/10.1016/j.ast.2008.11.001
12. Amitay, M., Smith, D.R., Kibens, V., Parekh, D.E., Glezer, A.: Aerodynamic flow control over an unconventional airfoil using synthetic jet actuators. AIAA J. **39**(3), 361–370 (2001)
13. Amitay, M., Pitt, D., Kibens, V., Parekh, D.E., Glezer, A.: Control of internal flow separation using synthetic jet actuators. AIAA, paper No. 2000-0903 (2000)
14. Amitay, M., Horvath, M., Michaux, M., Glezer, A.: Virtual aerodynamic shape modification at low angles of attack using synthetic jet actuators. AIAA, paper No. 2001-2975 (2001)
15. Chen, S.H., Tao, G., Fei, J.T., Joshi, S.M.: An adaptive actuator failure compensation scheme for controlling a morphing aircraft model. In: Proceedings of 42th IEEE Conference on Decision and Control, Maui, HI, pp. 4926–4931 (2003)
16. Crook, A., Wood, N.J: Measurements and Visualizations of Synthetic Jets Actuators, AIAA, paper No. 2001-0145, Jan 2001
17. Deb, D., Tao, G., Burkholder, J., Smith, D.: An adaptive inverse control scheme for a synthetic jet actuator model. Proc. Am. Control Conf. (2005). https://doi.org/10.1109/acc.2005.1470367
18. Gallas, Q., Holman, R., Nishida, T., Carroll, B., Sheplak, M., Cattafesta, L.: Lumped element modeling of Piezoelectric-Driven synthetic jet actuators, AIAA paper No. 2002-125, 14–17 January 2002
19. Beratlis, N., Smith, M.K.: Optimization of synthetic jet cooling for microelectronics applications. In: 19th Annual IEEE Semiconductor Thermal Measurement and Management Symposium, pp. 66–73, 11–13 Mar 2003
20. Honohan, A., Amitay, M., Glezer, A.: Aerodynamic control using synthetic jets, AIAA, paper No. 2000-2401 (2000)
21. Lewis, F.L.: Applied Optimal Control and Estimation. Prentice-Hall, Englewood Cliffs, NJ (1992)

22. Lockerby, D.: Numerical simulation of boundary-layer control using MEMS actuation, Ph.D Dissertation, The University of Warwick, Mar 2001
23. Raney, D.L., Montgomery, R.C., Green, L.L., Park, M.A.: Flight Control using Distributed Shape-Change Effector Arrays. AIAA paper No. 2000–1560, 3–6 Apr 2000
24. Rugh, W.J.: Linear System Theory, 2nd edn. Prentice-Hall, Englewood Cliffs, NJ (1996)
25. Smith, D., Amitay, M., Kibens, V., Parekh, D., Glezer, A.: Modification of the Lifting Body Aerodynamics by Synthetic Jet Actuators, AIAA, paper No. 98-0209 (1998)
26. Tao, G.: Adaptive Control Design and Analysis. Wiley (2003)
27. Tao, G., Kokotović, P.V.: Adaptive Control of Systems with Actuator and Sensor Nonlinearities. Wiley, New York (1996)
28. Arshad, A., Jabbal, M., Yan, Y.: Synthetic jet actuators for heat transfer enhancement—a critical review. Int. J. Heat Mass Transf. **146**, 118815. Elsevier BV (2020). https://doi.org/10.1016/j.ijheatmasstransfer.2019.118815

Chapter 3
Adaptive Compensation at Low Angles of Attack: Nonlinear Aircraft Model

The challenges encountered in the design of nonlinear control systems are very different from those faced by the equivalent linear systems. Recent work on two different types of nonlinear control systems like microbial fuel cells [1] and artificial pancreas [2] with challenging problems related to parametric uncertainties have evoked keen interest in related areas. Especially with SJA parametric uncertainties, the challenge is more as the uncertainty compensation has to be carried forth across all the control loops. When we extend the study of SJA compensation to nonlinear aircraft dynamical systems, we must consider that the number of available inputs is not sufficient to control all the states. Therefore, we utilize the available control signals to control the intermediate states, and those are then useful in the control of the other remaining state variables. We design the corresponding control signals, and an inverse compensator to account for the unknown actuator nonlinearities.

Ultimately, we design a feedback control law with intermediate states to control the subsequent aircraft states, along with other control signals like the engine thrust and the SJA generated lift forces. We design parameter adaptive update signals in the form of ordinary differential equations for the analysis of aircraft performance vis-a-vis closed-loop signal boundedness and asymptotic output tracking. Simulation results for the control of a nonlinear aircraft model verify the desired performance of our adaptive designs. The control law development is valid only at low AoA because the structure of the actuator nonlinearity, as will be shown in subsequent chapters, is modified at higher AoA.

The control objectives of this chapter are to develop:

(i) a nominal controller based on partial state feedback for a aircraft with mechanical flaps (or with nonlinearity compensated SJAs) through a comparative study of these two cases using simulation results. A a meaningful control allocation is also to be developed.

D. Deb et al., *Adaptive Compensation of Nonlinear Actuators for Flight Control Applications*, Studies in Systems, Decision and Control 386,
https://doi.org/10.1007/978-981-16-4161-9_3

(ii) an adaptive compensator (along with a simulation-based comparative study with a non-adaptive framework) to address the undesirable effects in certain aircraft states caused by nonlinearities in the SJs and for desired system performance.

3.1 Nonlinear Aircraft System: A Benchmark Model

We present a nonlinear aircraft model which is symmetric along the longitudinal axis and also in mass. The aerodynamic moments and forces depend on the aircraft orientation concerning the SJ flow. For simplified equations, we make the following assumptions for a generic nonlinear model:

(i) The earth is in rest and is considered to be flat which implies that any reference frame fixed to the earth (position and orientation) is an inertial reference system.
(ii) The equations are valid for a rigid body, that is, we neglect all elastic degrees of freedom of any part of the aircraft. The eigen-frequencies of the elastic modes are much higher than those of the rigid body motions.
(iii) The aircraft mass is constant.

The states of the nonlinear aircraft system are $x = \begin{bmatrix} p & q & r & \gamma & \chi & \mu & V & \alpha & \beta \end{bmatrix}^T$, where p, q, r (roll, pitch, and yaw, respectively) are the body-axis aircraft angular velocity components, V is the aircraft velocity magnitude, α is the angle of attack, β is the sideslip angle, γ, χ are the flight-path angles which are positive when the aircraft is climbing, and μ is the aircraft bank angle. These aircraft coordinates are detailed in [3]. The BANTAM configuration for operation under a wide range of AoA (both low and high AoA) consists by arrays of SJAs (six shaded regions) with boundary layer trips located forward in the air-stream, and a leading-edge jet array is utilized. The jets at the trailing edge are activated at low AoA [4], and the leading edge arrays at high AoA. At intermediate angles, one possible approach is to optimally select some jets from both the leading and trailing edges of the wing.

The objective is to formulate an unified control signal $u_d(t)$ for the reasonable AoA until stall. The equations of motion for air vehicles, typically include the side-force and drag-force terms [5, 6]. However, in this study, the side forces are negligible because the type of SJs in use actuates in a direction normal to the airfoil surface. With the provision for surface-mounted miniature passive trips, the localized drag forces are reasonable to neglect in the present analysis. Therefore, as far as the aerodynamic forces are concerned, we focus entirely on the lift forces generated by the SJAAs in the six different regions of the aircraft under consideration in this study. The control inputs are the thrust signal T, and the lift force u generated by SJAs is given by

$$u(t) = \sum_{i=1}^{6} u_i(t),$$

where $u_i(t), i = 1, \dots 6$ represents the consolidated lift forces as a result of actuation of the six SJ arrays. More details about individual actuators are available in literature [7]. We make an assumption that the outputs u_i are not measurable and only the parameter bounds of $N_i(\cdot)$ are known.

With the rotational inertia parameters $c_i, i = 1, \dots, 7$ expressed as

$$c_1 = \frac{I_{yy} - I_{zz}}{I_{xx}}, \; c_2 = \frac{I_{xx} - I_{yy} + I_{zz}}{I_{xx} I_{zz}},$$

$$c_3 = \frac{1}{I_{xx}}, \; c_4 = \frac{I_{zz} - I_{xx}}{I_{yy}},$$

$$c_5 = \frac{1}{I_{yy}}, \; c_6 = \frac{I_{xx} - I_{yy}}{I_{zz}}, \; c_7 = \frac{I_{xx}}{I_{xx} I_{zz} - I_{xz}^2},$$

with $I_{ij}, \; i = x, y, z, j = x, y, z$, and $I_{yz} = I_{yx} = I_{xz} = 0$, the nonlinear dynamical equations are given as

$$\dot{V} = -g \, \text{sn}\gamma + \frac{T}{m} \text{cs}\alpha \, \text{cs}\beta, \tag{3.1}$$

$$\dot{\mu} = \frac{p \, \text{cs}\alpha + r \, \text{sn}\, \alpha}{\text{cs}\beta} + \frac{1}{mV} (\text{tn}\, \beta + \text{tn}\, \gamma \, \text{sn}\, \mu) \, u + \frac{T}{mV} (\text{sn}\, \alpha \, \text{sn}\, \mu \, \text{tn}\, \gamma$$
$$- \text{cs}\alpha \, \text{sn}\, \beta \, \text{tn}\, \gamma \, \text{cs}\mu + \text{sn}\, \alpha \, \text{tn}\, \beta)) - \frac{g}{V} \text{tn}\, \beta \, \text{cs}\gamma \, \text{cs}\mu,$$

$$\dot{\alpha} = -\frac{1}{mV \, \text{cs}\beta}(u - mg \, \text{cs}\gamma \, \text{cs}\mu + T \, \text{sn}\, \alpha) - p \, \text{cs}\alpha \, \text{tn}\, \beta + q - r \, \text{sn}\, \alpha \, \text{tn}\, \beta, \tag{3.2}$$

$$\dot{\beta} = \frac{1}{mV}(mg \, \text{sn}\, \mu \, \text{cs}\gamma - T \, \text{cs}\alpha \, \text{sn}\, \beta) + p \, \text{sn}\, \alpha - r \, \text{cs}\alpha, \tag{3.3}$$

$$\dot{\gamma} = \frac{1}{mV} (u \, \text{cs}\mu + T(\text{sn}\, \alpha \, \text{cs}\mu + \text{cs}\alpha \, \text{sn}\, \beta \, \text{sn}\, \mu) - mg \, \text{cs}\gamma) , \tag{3.4}$$

$$\dot{\chi} = \frac{1}{mV \, \text{cs}\gamma} (u \, \text{sn}\, \mu - T(\text{cs}\alpha \, \text{sn}\, \beta \, \text{cs}\mu - \text{sn}\, \alpha \, \text{sn}\, \mu)), \tag{3.5}$$

$$\dot{p} = (c_1 r + c_2 p)q + c_3 \bar{l}, \tag{3.6}$$

$$\dot{q} = c_4 pr + c_5 \bar{m}, \tag{3.7}$$

$$\dot{r} = (c_6 p - c_2 r)q + c_7 \bar{n}, \tag{3.8}$$

where sn, cs and tn represent sine, cosine and tangent functions respectively, T is the thrust, u is the airfoil lift force, m is the mass, and $\bar{l}, \bar{m}, \bar{n}$ are the roll, pitch and yaw moments respectively [8].

An adaptive compensator for the nonlinearities $N(\cdot)$ is shown in Fig. 3.1 where $\dot{x} = f_x + g_{x,u}$ represents the aircraft dynamics, $N(\cdot) = [N_1(\cdot), \dots, N_6(\cdot)]^T$ and $\left[\widehat{NI}_1(\cdot), \dots, \widehat{NI}_6(\cdot)\right]^T$ are the SJAA uncertainties and the corresponding adaptive inversion functions respectively, and $u_{di}(t), i = 1, \dots, 6$, are the desired lift forces.

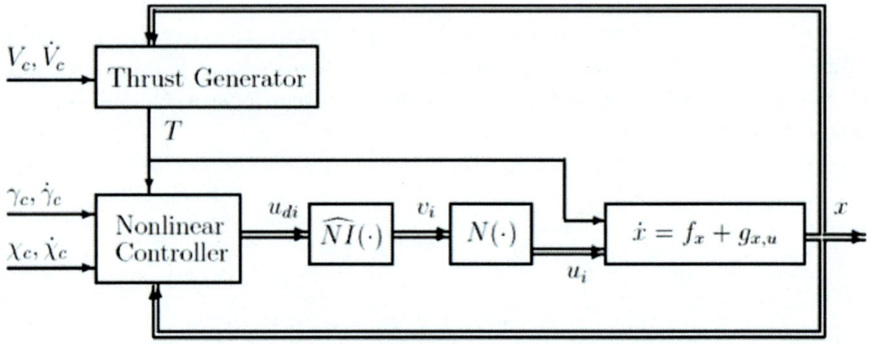

Fig. 3.1 Adaptive actuator compensation scheme

The thrust T is a control signal obtained by feedback along with the desired lift signals $u_{di}(t)$, V_r, γ_r, ξ_r (and derivatives) are externally available references that are known to be bounded. We can appropriately use any available nominal flight controller as the baseline. The adaptive inverse control technique is augmented with such a controller to accommodate the SJAs without the need for extensive wind tunnel tests.

Next, we generate an adaptive control signal from accessible intermediate states as control laws to be used in the control action of other states. Lift forces acting on the airfoil due to the SJAs and the engine thrust are the available control options. The adaptive inverse-based approach at low AoA which generates u_i, is not valid at higher AoA because the actuator nonlinearity gets modified leading to the need for a different nonlinear controller as presented in the last three chapters of this book. Note that although the thrust signal T would also change at different AoA, but the change is not dependent on actuator uncertainties.

3.2 Feedback Control Law Design

In this formulation, there are three control loops. The action of external inputs on certain states causes those states to be used as control signals for subsequent states. The desired lift forces $u_{di}, i = 1, \ldots, 6$ control the body axis rates p, q, r, and the thrust T control the velocity V. The other states such as β are regulated to 0, and μ, α control χ, γ. The states p, q, r are used to control the signals μ, α and β. The definition of the aircraft body axis states are shown in Fig. 3.2.

The control law formulation evolves from the iterative control concept [9]. We include new details like thrust signal generation for negligible drag, solvability challenges, and allocation of desired lift forces so as to incorporate aerodynamic changes and accomplish stability considerations. Approximation conditions for higher-order terms [10] are also described herewith. A robust controller is designed that does not require switching or high gain.

Fig. 3.2 Definition of aircraft body axis states

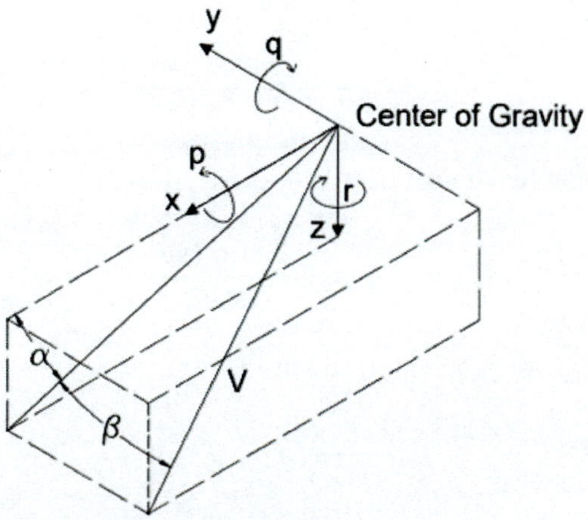

We make the following reasonable assumption:

(A2) the velocity V is not close to 0, and the angles α, β, γ are not equal to 90 degrees to avoid singularity situations.

3.2.1 *Velocity Control*

Using the dynamical equation pertaining to velocity (3.1), for achieving control of velocity, an external signal which is not from the actuators, the thrust T can be defined as

$$T = \frac{m}{\mathrm{cs}\alpha\, \mathrm{cs}\beta}\left(g\, \mathrm{sn}\, \gamma + \dot{V}_r - K_V \tilde{V}\right), \tag{3.9}$$

for a reference signal V_r, and velocity error $\tilde{V} = V - V_r$, and gain $K_V > 0$.

Using (3.1), (3.9), the velocity error dynamics is

$$\dot{\tilde{V}} = -K_V \tilde{V}. \tag{3.10}$$

3.2.2 *Flight Path Angle Dynamics*

With thrust T (3.9), the flight path angle dynamics (3.4) and (3.5) can be given by

$$\dot{\gamma} = f_\gamma + g_{\gamma,u}, \quad \dot{\chi} = f_\chi + g_{\chi,u}, \tag{3.11}$$

where the known functions are

$$f_\gamma = \frac{-mg\,\mathrm{cs}\gamma + T\,\mathrm{cs}\alpha\,\mathrm{sn}\,\beta\,\mathrm{sn}\,\mu}{mV}, \quad f_\chi = -\frac{T\,\mathrm{cs}\alpha\,\mathrm{sn}\,\beta\,\mathrm{cs}\mu}{mV\,\mathrm{cs}\gamma},$$

the functions $g_{\gamma,u}$ and $g_{\chi,u}$ are given by

$$g_{\gamma,u} = \frac{\mathrm{cs}\mu}{mV}u + \frac{T\,\mathrm{sn}\,\alpha\,\mathrm{cs}\mu}{mV}, \quad g_{\chi,u} = \frac{\mathrm{sn}\,\mu}{mV\,\mathrm{cs}\gamma}u + \frac{T\,\mathrm{sn}\,\alpha\,\mathrm{sn}\,\mu}{mV\,\mathrm{cs}\gamma}. \tag{3.12}$$

We generate μ_r, α_r from

$$\begin{aligned} \frac{\mathrm{cs}\mu_r}{mV}(u_d + T\,\mathrm{sn}\,\alpha_r) &= \dot{\gamma}_r - f_\gamma - K_\gamma\tilde{\gamma}, \;\; K_\gamma > 0, \\ \frac{\mathrm{sn}\,\mu_r}{mV\,\mathrm{cs}\gamma}(u_d + T\,\mathrm{sn}\,\alpha_r) &= \dot{\chi}_r - f_\chi - K_\chi\tilde{\chi}, \;\; K_\chi > 0, \end{aligned} \tag{3.13}$$

where χ_r and γ_r are the reference flight path angles, $\tilde{\gamma} = \gamma - \gamma_r$, $\tilde{\chi} = \chi - \chi_r$, and μ_r, α_r are the controls in the wind axis loop.

From (3.13), μ_r and α_r are given by

$$\mu_r = \arctan\left(\frac{\mathrm{cs}\gamma\,(\dot{\chi}_r - f_\chi - K_\chi\tilde{\chi})}{\dot{\gamma}_r - f_\gamma - K_\gamma\tilde{\gamma}}\right), \quad \alpha_r = \arcsin\left(\frac{mVF_\gamma - u_d}{T}\right) \tag{3.14}$$

where $F_\gamma = \frac{\dot{\gamma}_r - f_\gamma - K_\gamma\tilde{\gamma}}{\mathrm{cs}\mu_r}$. The validity of α_r requires

$$\left|mVF_\gamma - u_d\right| < \kappa\,|T|, \tag{3.15}$$

where the range $\alpha_r \in (-25°, 25°)$ ensures allowable range for $\kappa = 0.1736$. Factors such as the input voltage and the number of SJAs deployed determine whether sufficient lift force can be generated so as to satisfy (3.15).

Using (3.11)–(3.13), the rate of change of $\tilde{\gamma}, \tilde{\chi}$ are

$$\begin{aligned} \dot{\tilde{\gamma}} &= -K_\gamma\tilde{\gamma} + \frac{T\,(\mathrm{sn}\,\alpha\,\mathrm{cs}\mu - \mathrm{sn}\,\alpha_r\,\mathrm{cs}\mu_r)}{mV} + \frac{\mathrm{cs}\mu}{mV}u - \frac{\mathrm{cs}\,\mu_r}{mV}u_d, \\ \dot{\tilde{\chi}} &= -K_\chi\tilde{\chi} + \frac{T\,(\mathrm{sn}\,\alpha\,\mathrm{sn}\,\mu - \mathrm{sn}\,\alpha_r\,\mathrm{sn}\,\mu_r)}{mV\,\mathrm{cs}\gamma} + \frac{\mathrm{sn}\,\mu}{mV\,\mathrm{cs}\gamma}u - \frac{\mathrm{sn}\,\mu_r}{mV\,\mathrm{cs}\gamma}u_d. \end{aligned} \tag{3.16}$$

Using Taylor series expansions, we get

$$\begin{aligned} \mathrm{cs}\mu &= \mathrm{cs}\mu_r + O_1(\tilde{\mu}), \quad \mathrm{sn}\,\mu = \mathrm{sn}\,\mu_r + O_2(\tilde{\mu}), \\ \mathrm{sn}\,\alpha\,\mathrm{cs}\mu &= \mathrm{sn}\,\alpha_r\,\mathrm{cs}\mu_r + \tilde{\alpha}\,\mathrm{cs}\alpha_r\,\mathrm{cs}\mu_r - \tilde{\mu}\,\mathrm{sn}\,\alpha_r\,\mathrm{sn}\,\mu_r + O_3(\tilde{\mu}, \tilde{\alpha}), \\ \mathrm{sn}\,\alpha\,\mathrm{sn}\,\mu &= \mathrm{sn}\,\alpha_r\,\mathrm{sn}\,\mu_r + \tilde{\mu}\,\mathrm{sn}\,\alpha_r\,\mathrm{cs}\mu_r + \tilde{\alpha}\,\mathrm{sn}\,\mu_r\,\mathrm{cs}\alpha_r + O_4(\tilde{\mu}, \tilde{\alpha}), \end{aligned} \tag{3.17}$$

where $\tilde{\mu} = \mu - \mu_r, \tilde{\alpha} = \alpha - \alpha_r$, and $O_i(\tilde{\mu}), i = 1, 2,$ and $O_i(\tilde{\mu}, \tilde{\alpha}), i = 3, 4,$ are composed of second and higher order components of $\tilde{\mu}$:

$$\lim_{|\tilde{\mu}| \to 0} \frac{O_i(\tilde{\mu})}{|\tilde{\mu}|} = 0, i = 1, 2,$$

$$\lim_{|\tilde{\mu}|, |\tilde{\alpha}| \to 0} \frac{O_i(\tilde{\mu}, \tilde{\alpha})}{\sqrt{|\tilde{\mu}|^2 + |\tilde{\alpha}|^2}} = 0, i = 3, 4.$$

Using (3.16) and (3.17), the approximated flight path angle errors are given by

$$\begin{aligned}
\dot{\tilde{\gamma}} &= -K_\gamma \tilde{\gamma} + \frac{\mathrm{cs}\mu}{mV}(u - u_d) + \frac{T\,(\tilde{\alpha}\,\mathrm{cs}\alpha_r\,\mathrm{cs}\mu_r - \tilde{\mu}\,\mathrm{sn}\,\mu_r\,\mathrm{sn}\,\alpha_r)}{mV} \\
&+ \frac{O_1(\tilde{\mu})u_d + O_3(\tilde{\mu}, \tilde{\alpha})T}{mV}, \\
\dot{\tilde{\chi}} &= -K_\chi \tilde{\chi} + \frac{\mathrm{sn}\,\mu}{mV\,\mathrm{cs}\gamma}(u - u_d) + \frac{T\,(\tilde{\alpha}\,\mathrm{sn}\,\mu_r\,\mathrm{cs}\alpha_r + \tilde{\mu}\,\mathrm{cs}\mu_r\,\mathrm{sn}\,\alpha_r)}{mV\,\mathrm{cs}\gamma} \\
&+ \frac{O_2(\tilde{\mu})u_d + O_4(\tilde{\mu}, \tilde{\alpha})T}{mV\,\mathrm{cs}\gamma}.
\end{aligned} \tag{3.18}$$

3.2.3 *Wind Axis Angle Dynamics*

The wind axis angle dynamics (3.2), (3.3) are given as

$$\dot{\mu} = f_\mu + g_{\mu,u}, \quad \dot{\alpha} = f_\alpha + g_{\alpha,u}, \quad \dot{\beta} = f_\beta + g_{\beta,u}, \tag{3.19}$$

for known functions f_μ, f_α and f_β are

$$\begin{aligned}
f_\mu &= \frac{T}{mV}(\mathrm{sn}\,\alpha\,\mathrm{tn}\,\gamma\,\mathrm{sn}\,\mu + \mathrm{sn}\,\alpha\,\mathrm{tn}\,\beta - \mathrm{cs}\alpha\,\mathrm{tn}\,\gamma\,\mathrm{cs}\mu\,\mathrm{sn}\,\beta) - \frac{g}{V}\mathrm{cs}\alpha\,\mathrm{cs}\mu\,\mathrm{tn}\,\beta, \\
f_\alpha &= \frac{1}{mV\,\mathrm{cs}\beta}(-T\,\mathrm{sn}\,\alpha + mg\,\mathrm{cs}\gamma\,\mathrm{cs}\mu) - (p\,\mathrm{cs}\alpha + r\,\mathrm{sn}\,\alpha)\,\mathrm{tn}\,\beta, \\
f_\beta &= \frac{1}{mV}(-T\,\mathrm{sn}\,\beta\,\mathrm{cs}\alpha + mg\,\mathrm{cs}\gamma\,\mathrm{sn}\,\mu),
\end{aligned} \tag{3.20}$$

and $g_{\mu,u}$, $g_{\alpha,u}$ and $g_{\beta,u}$ are

$$\begin{aligned}
g_{\mu,u} &= \frac{p\,\mathrm{cs}\alpha + r\,\mathrm{sn}\,\alpha}{\mathrm{cs}\beta} + \frac{\mathrm{tn}\,\beta + \mathrm{tn}\,\gamma\,\mathrm{sn}\,\mu}{mV}u, \\
g_{\alpha,u} &= -\frac{1}{mV\,\mathrm{cs}\beta}u + q, \quad g_{\beta,u} = p\,\mathrm{sn}\,\alpha - r\,\mathrm{cs}\alpha.
\end{aligned} \tag{3.21}$$

For the control action on the wind axis angles control, we develop the appropriate signals p_r, q_r, r_r:

$$\frac{p_r \,\mathrm{cs}\alpha + r_r \,\mathrm{sn}\,\alpha}{\mathrm{cs}\beta} + \frac{\mathrm{tn}\,\beta + \mathrm{tn}\,\gamma\,\mathrm{sn}\,\mu}{mV} u_d = -f_\mu + \dot{\mu}_r - K_\mu \tilde{\mu} + \frac{T\,\mathrm{sn}\,\alpha_r}{mV}\left(-\tilde{\chi}\frac{\mathrm{cs}\mu_r}{\mathrm{cs}\gamma} + \tilde{\gamma}\,\mathrm{sn}\,\mu_r\right),$$

$$q_r - \frac{1}{mV\,\mathrm{cs}\beta} u_d = -f_\alpha + \dot{\alpha}_r - K_\alpha \tilde{\alpha} + \frac{T\,\mathrm{cs}\alpha_r}{mV}\left(-\tilde{\chi}\frac{\mathrm{sn}\,\mu_r}{\mathrm{cs}\gamma} - \tilde{\gamma}\,\mathrm{cs}\mu_r\right),$$
$$p_r \,\mathrm{sn}\,\alpha - r_r\,\mathrm{cs}\alpha = -f_\beta + \dot{\beta}_r - K_\beta \tilde{\beta}, \tag{3.22}$$

where μ_r, α_r are obtained using (3.13), and available for this loop, and $K_\mu, K_\alpha, K_\beta > 0$ are constant gains, $\tilde{\mu} = \mu - \mu_r$, $\tilde{\alpha} = \alpha - \alpha_r$, $\tilde{\beta} = \beta - \beta_r$ are the tracking errors. For the desired references to be achievable, the following bandwidth related rules must be followed:

$$K_p > K_\mu > K_\chi, \quad K_q > K_\alpha > K_\gamma, \quad K_r > K_\beta. \tag{3.23}$$

Using (3.19)–(3.22), the dynamic equations for $\tilde{\mu} = \mu - \mu_r, \tilde{\alpha} = \alpha - \alpha_r$ and $\tilde{\beta} = \beta - \beta_r$, are given by

$$\dot{\tilde{\mu}} = -K_\mu \tilde{\mu} - \frac{T\,\mathrm{sn}\,\alpha_r}{mV}\left(\tilde{\chi}\frac{\mathrm{cs}\mu_r}{\mathrm{cs}\gamma} - \tilde{\gamma}\,\mathrm{sn}\,\mu_r\right) + \frac{\tilde{p}\,\mathrm{cs}\alpha + \tilde{r}\,\mathrm{sn}\,\alpha}{\mathrm{cs}\beta} + \frac{\mathrm{tn}\,\beta + \mathrm{tn}\,\gamma\,\mathrm{sn}\,\mu}{mV}(u - u_d),$$
$$\dot{\tilde{\alpha}} = -K_\alpha \tilde{\alpha} + \tilde{q} + \frac{T\,\mathrm{cs}\alpha_r}{mV}\left(-\tilde{\chi}\frac{\mathrm{sn}\,\mu_r}{\mathrm{cs}\gamma} - \tilde{\gamma}\,\mathrm{cs}\mu_r\right) - \frac{1}{mV\,\mathrm{cs}\beta}(u - u_d),$$
$$\dot{\tilde{\beta}} = -K_\beta \tilde{\beta} + \tilde{p}\,\mathrm{sn}\,\alpha - \tilde{r}\,\mathrm{cs}\alpha, \tag{3.24}$$

for body axis rate errors $\tilde{p} = p - p_r, \tilde{q} = q - q_r, \tilde{r} = r - r_r$.

3.2.4 Body Axis Rate Dynamics

The body axis equations given by (3.6)–(3.8) are

$$\dot{p} = f_p + g_{p,u}, \quad \dot{q} = f_q + g_{q,u}, \quad \dot{r} = f_r + g_{r,u}, \tag{3.25}$$

where the known functions for known aerodynamic moments $\bar{l}_a(t), \bar{m}_a(t), \bar{n}_a(t)$ are

$$\begin{aligned} f_p &= (c_1 r + c_2 p)q + c_3 \bar{l}_a(t), \\ f_q &= c_4 pr + c_5 \bar{m}_a(t), \\ f_r &= (c_6 p - c_2 r)q + c_7 \bar{l}_a(t), \end{aligned} \tag{3.26}$$

and the functions $g_{p,u}, g_{q,u}, g_{r,u}$ are given as

$$g_{p,u} = c_3 \sum_{i=1}^{6} a_{1i}(t) u_i(t), \;\; g_{q,u} = c_5 \sum_{i=1}^{6} a_{2i}(t) u_i(t), \;\; g_{r,u} = c_7 \sum_{i=1}^{6} a_{3i}(t) u_i(t).$$

where the stability axis roll, pitch and yaw moments are $\sum_{i=1}^{6} a_{1i}(t)u_i(t) = \bar{l}(t) - \bar{l}_a(t)$, $\sum_{i=1}^{6} a_{2i}(t)u_i(t) = \bar{m}(t) - \bar{m}_a(t)$, $\sum_{i=1}^{6} a_{3i}(t)u_i(t) = \bar{n}(t) - \bar{n}_a(t)$ respectively, and $a_{1i}, a_{2i}, a_{3i}, i = 1, \ldots, 6$ are the functional weights that determine the quantum of desired lift force allocation due to u_i.

The desired signals $u_{di}(t), i = 1, \ldots, 6$, are obtained from

$$\begin{aligned} c_3 \sum_{i=1}^{6} a_{1i}(t) u_{di}(t) &= -f_p - K_p \tilde{p} + \dot{p}_r - \left(\tilde{\mu} \frac{\mathrm{cs}\alpha}{\mathrm{cs}\beta} + \tilde{\beta}\, \mathrm{sn}\, \alpha\right), \; K_p > 0, \\ c_5 \sum_{i=1}^{6} a_{2i}(t) u_{di}(t) &= -f_q + \dot{q}_r - K_q \tilde{q} - \tilde{\alpha}, \; K_q > 0, \\ c_7 \sum_{i=1}^{6} a_{3i}(t) u_{di}(t) &= -f_r - K_r \tilde{r} + \dot{r}_r - \left(\tilde{\mu} \frac{\mathrm{sn}\, \alpha}{\mathrm{cs}\beta} - \tilde{\beta}\, \mathrm{cs}\alpha\right), \; K_r > 0, \end{aligned} \tag{3.27}$$

where p_r, q_r, r_r obtained from (3.22) are available for body axis rate control.

In (3.27), some functions containing error terms $\tilde{\mu}, \tilde{\alpha}, \tilde{\beta}$, are inserted for ensuring Lyapunov stability. Because of the symmetry of the SJ arrays vis-a-vis the longitudinal axis, the outer trailing edge jet arrays, the middle trailing edge jet arrays, and the spoiler jet arrays significantly affect the respective moments [11]. A possible allocation policy for control authority is given herewith:

$$u_{d1}(t) = -\frac{a_{14}(t)}{a_{11}(t)} u_{d4}(t), \;\; u_{d2}(t) = -\frac{a_{25}(t)}{a_{22}(t)} u_{d5}(t), \;\; u_{d3}(t) = -\frac{a_{36}(t)}{a_{33}(t)} u_{d6}(t), \tag{3.28}$$

Such a control allocation reduces (3.27) into a form

$$\begin{bmatrix} u_{d1} \\ u_{d2} \\ u_{d3} \end{bmatrix} = A^{-1} \begin{bmatrix} \frac{1}{c_3}\left(-f_p - K_p \tilde{p} + \dot{p}_r - \left(\tilde{\mu} \frac{\mathrm{cs}\alpha}{\mathrm{cs}\beta} + \tilde{\beta}\, \mathrm{sn}\, \alpha\right)\right) \\ \frac{1}{c_5}\left(-f_q + \dot{q}_r - K_q \tilde{q} - \tilde{\alpha}\right) \\ \frac{1}{c_7}\left(-f_r - K_r \tilde{r} + \dot{r}_r - \left(\tilde{\mu} \frac{\mathrm{sn}\, \alpha}{\mathrm{cs}\beta} - \tilde{\beta}\, \mathrm{cs}\alpha\right)\right) \end{bmatrix}, \tag{3.29}$$

where the matrix A is given by

$$A = \begin{bmatrix} 0 & a_{12} + \frac{a_{15}a_{22}}{a_{25}} & a_{13} - \frac{a_{16}a_{33}}{a_{36}} \\ a_{21} - \frac{a_{24}a_{11}}{a_{14}} & 2a_{22} & a_{23} - \frac{a_{26}a_{33}}{a_{36}} \\ a_{31} - \frac{a_{34}a_{11}}{a_{14}} & a_{32} + \frac{a_{35}a_{22}}{a_{25}} & 0 \end{bmatrix}. \tag{3.30}$$

For a non-singular matrix A, the desired control laws $u_{di}(t), i = 1, 2, 3$ are obtained from (3.29) and so from (3.28), $u_{di}(t), i = 4, 5, 6$ follows. The non-singularity of matrix A depends on the values $a_{ij}, i = 1, 2, 3, j = 1, \ldots, 6$.

Using (3.25) and (3.27), the rate of $\tilde{p} = p - p_r, \tilde{q} = q - q_r, \tilde{r} = r - r_r$, are

$$\begin{aligned} \dot{\tilde{p}} &= -\left(\tilde{\mu}\frac{\text{cs}\alpha}{\text{cs}\beta} + \tilde{\beta}\,\text{sn}\,\alpha\right) + c_3 a_1 (u - u_d) - K_p\tilde{p}, \\ \dot{\tilde{q}} &= c_5 a_2 (u - u_d) - K_q\tilde{q} - \tilde{\alpha}, \\ \dot{\tilde{r}} &= \tilde{\beta}\,\text{cs}\alpha - \tilde{\mu}\frac{\text{sn}\,\alpha}{\text{cs}\beta} - K_r\tilde{r} + c_7 a_3 (u - u_d)\,. \end{aligned} \tag{3.31}$$

3.3 Performance Analysis

The parameters $\theta_{1i}(t)$ and $\theta_{2i}(t)$ must remain bounded within certain desired values for appropriate adaptation. Parameter projection design ensures that the estimates $\theta_{1i}(t)$ and $\theta_{2i}(t)$ stay in the same region where their true values θ^*_{1i} and θ^*_{2i} stay and their lower bounds are known $0 < \theta^a_{1i} \leq \theta^*_{1i}$ and $0 < \theta^a_{2i} \leq \theta^*_{2i}$ [12, 13]. Other robust adaptation techniques can also be used [14].

Initial estimates $\theta_{1i}(0)$ and $\theta_{2i}(0)$ of θ^*_{1i} and θ^*_{2i} respectively, are chosen such that

$$\theta_{1i}(t) > \theta^a_{1i}, \quad \theta_{2i}(t) > \theta^a_{2i}. \tag{3.32}$$

We formulate the estimated update laws as

$$\dot{\theta}_{ji}(t) = g_{ji}(t) + f_{ji}(t), \; i = 1, \ldots, 6, \; j = 1, 2, \tag{3.33}$$

with projections $f_{ji}(t)$ expressed as

$$f_{ji}(t) = \begin{cases} 0 & \text{if } \; \theta_{ji}(t) > \theta^a_{ji}, \; \text{ or} \\ & \text{if } \; \theta_{ji}(t) = \theta^a_{ji} \text{ and } \; g_{ji}(t) \geq 0, \; j = 1, 2, \; i = 1, \ldots, 6 \\ -g_{ji}(t) & \text{otherwise.} \end{cases} \tag{3.34}$$

and the functions $g_{ji}(t)$ are expressed as

$$g_{1i} = -\gamma_{1i}\,(e(t) + \bar{e}_i(t))\left(\frac{\theta_{2i}(t) - u_{di}(t)}{\theta_{1i}}\right),$$
$$g_{2i} = \gamma_{2i}\,(e(t) + \bar{e}_i(t)),\ \gamma_{1i}, \gamma_{2i} > 0,\ i = 1, \ldots, 6, \tag{3.35}$$

and a specifically constructed error function expressed as

$$\bar{e}_i(t) = c_3\tilde{p}(t)a_{1i}(t) + c_5\tilde{q}(t)a_{2i}(t) + c_7\tilde{r}(t)a_{3i}(t),$$
$$e(t) = \frac{1}{mV}\left(-\frac{\tilde{\chi}\,\mathrm{sn}\,\mu}{\mathrm{cs}\gamma} - \tilde{\gamma}\,\mathrm{cs}\mu + \frac{\tilde{\alpha}}{\mathrm{cs}\beta} + \tilde{\mu}\,(\mathrm{tn}\,\beta + \mathrm{tn}\,\gamma\,\mathrm{sn}\,\mu)\right). \tag{3.36}$$

Note that the terms in $\bar{e}_i(t), e(t)$ are introduced in a way that certain errors in the Eqs. (3.18), (3.24), and (3.31) are eventually canceled out during stability analysis which is presented in the next section. This choice of the adaptive laws has the result that the time derivative of the Lyapunov function (of the following section) is negative semi-definite. The Lyapunov stability analysis demonstrates that the size of a small of the origin neighborhood is determined (in part) by the residual approximation error.

Theorem 3.1 *Under assumptions (A1), (A2), and small errors $\tilde{\mu}, \tilde{\alpha}$, the tracking errors (3.10), (3.18), (3.24), and (3.31) from feedback control signals, with adaptively updated parameters from (3.35), guarantee signal boundedness and imply that $\tilde{x}(t)$ is small [15, 16].*

Proof Let us consider a scalar continuous and positive definite expression

$$W = \frac{1}{2}\tilde{x}^T\tilde{x} + \frac{1}{2}\sum_{i=1}^{6}\sum_{j=1}^{2}\tilde{\theta}_{ji}^2\gamma_{ji}^{-1}, \tag{3.37}$$

as a measure of errors $\tilde{x}, \tilde{\theta}_{1i}(t), \tilde{\theta}_{2i}(t),\ i = 1, \ldots, 6$.

The time-derivative of W is

$$\dot{W} = \tilde{x}^T\dot{x} + \frac{1}{2}\sum_{i=1}^{6}\sum_{j=1}^{2}\tilde{\theta}_{ji}\gamma_{ji}^{-1}\dot{\theta}_{ji}. \tag{3.38}$$

Using (3.10), (3.18), (3.24), and (3.31), we get

$$
\begin{aligned}
\dot{W} = & -K_v\tilde{V}^2 - K_\chi\tilde{\chi}^2 + \frac{T\tilde{\chi}\,(\tilde{\alpha}\,\mathrm{cs}\alpha_r\,\mathrm{sn}\,\mu_r + \tilde{\mu}\,\mathrm{sn}\,\alpha_r\,\mathrm{cs}\mu_r)}{mV\,\mathrm{cs}\gamma} + \frac{\tilde{\chi}\,\mathrm{sn}\,\mu}{mV\,\mathrm{cs}\gamma}(u - u_d) \\
& -K_\gamma\tilde{\gamma}^2 + \frac{T\tilde{\gamma}\,(\tilde{\alpha}\,\mathrm{cs}\alpha_r\,\mathrm{cs}\mu_r - \tilde{\mu}\,\mathrm{sn}\,\alpha_r\,\mathrm{sn}\,\mu_r)}{mV} + \frac{\tilde{\mu}\,(\tilde{p}\,\mathrm{cs}\alpha + \tilde{r}\,\mathrm{sn}\,\alpha)}{\mathrm{cs}\beta} + \tilde{q}\tilde{\alpha} \\
& -\frac{\tilde{\gamma}\,\mathrm{cs}\mu}{mV}(u - u_d) - k_\mu\tilde{\mu}^2 - \frac{T\tilde{\mu}\,\mathrm{sn}\,\alpha_r}{mV}\left(\tilde{\chi}\frac{\mathrm{cs}\mu_r}{\mathrm{cs}\gamma} - \tilde{\gamma}\,\mathrm{sn}\,\mu_r\right) - K_\alpha\tilde{\alpha}^2 \\
& +\frac{\tilde{\mu}}{mV}(\mathrm{tn}\,\beta + \mathrm{tn}\,\gamma\,\mathrm{sn}\,\mu)(u - u_d) + \frac{T\tilde{\alpha}\,\mathrm{cs}\alpha_r}{mV}\left(-\tilde{\chi}\frac{\mathrm{sn}\,\mu_r}{\mathrm{cs}\gamma} - \tilde{\gamma}\,\mathrm{cs}\mu_r\right) \\
& -K_\beta\tilde{\beta}^2 + \tilde{\beta}\,(\tilde{p}\,\mathrm{sn}\,\alpha - \tilde{r}\,\mathrm{cs}\alpha) + \frac{\tilde{\alpha}}{mV\,\mathrm{cs}\beta}(u - u_d) - \tilde{p}\left(\tilde{\mu}\frac{\mathrm{cs}\alpha}{\mathrm{cs}\beta} + \tilde{\beta}\,\mathrm{sn}\,\alpha\right) \\
& -K_p\tilde{p}^2 - K_q\tilde{q}^2 - K_r\tilde{r}^2 + (\tilde{p}c_3a_1 + \tilde{q}c_5a_2 + \tilde{r}c_7a_3)(u - u_d) - \tilde{q}\left(K_q\tilde{q} + \tilde{\alpha}\right) \\
& -\tilde{r}\left(\tilde{\mu}\frac{\mathrm{sn}\,\alpha}{\mathrm{cs}\beta} - \tilde{\beta}\,\mathrm{cs}\alpha\right) + \frac{1}{2}\sum_{i=1}^{6}\sum_{j=1}^{2}\tilde{\theta}_{ji}\gamma_{ji}^{-1}\dot{\theta}_{ji} + \Delta,
\end{aligned} \tag{3.39}
$$

where Δ represents the third order error terms given by

$$
\Delta = \frac{\tilde{\gamma}\,(O_1(\tilde{\mu})u_d + O_3(\tilde{\mu}, \tilde{\alpha})T)}{mV} + \frac{\tilde{\chi}\,(O_2(\tilde{\mu})u_d + O_4(\tilde{\mu}, \tilde{\alpha})T)}{mV\,\mathrm{cs}\gamma}, \tag{3.40}
$$

with $\tilde{\mu}, \tilde{\gamma}$ for flight path error dynamics (3.18), and $O_i(\tilde{\mu}), i = 1, 2,\ O_i(\tilde{\mu}, \tilde{\gamma}), i = 3, 4,$ representing higher order error terms such that

$$
\Delta \le \delta_1\tilde{\gamma} + \delta_2\tilde{\chi}, \tag{3.41}
$$

for computable bounds δ_1, δ_2. From (3.36), (3.37) and (3.41), and adaptive laws (3.33), (3.35), and through a careful common term cancellation process, we get

$$
\begin{aligned}
\dot{W} \le & -K_V\tilde{V}^2 - K_\gamma\tilde{\gamma}^2 - K_\chi\tilde{\chi}^2 - K_\mu\tilde{\mu}^2 - K_\beta\tilde{\beta}^2 - K_\alpha\tilde{\alpha}^2 - K_p\tilde{p}^2 - K_q\tilde{q}^2 - K_r\tilde{r}^2 \\
& +e(t)\,(u - u_d) + \delta_1\tilde{\gamma} + \delta_2\tilde{\chi} + \sum_{i=1}^{6}\bar{e}_i(t)\,(u_i - u_{di}) + \frac{1}{2}\sum_{i=1}^{6}\sum_{j=1}^{2}\tilde{\theta}_{ji}\gamma_{ji}^{-1}\dot{\theta}_{ji} \\
\le & -K_V\tilde{V}^2 - K_\gamma\tilde{\gamma}^2 - K_\chi\tilde{\chi}^2 - K_\mu\tilde{\mu}^2 - K_\beta\tilde{\beta}^2 - K_\alpha\tilde{\alpha}^2 - K_p\tilde{p}^2 - K_q\tilde{q}^2 - K_r\tilde{r}^2 \\
& +\delta_1\tilde{\gamma} + \delta_2\tilde{\chi} + \frac{1}{2}\sum_{i=1}^{6}\sum_{j=1}^{2}\tilde{\theta}_{ji}\gamma_{ji}^{-1}f_{ji}.
\end{aligned} \tag{3.42}
$$

Using the condition $\theta_{ji}(0) > \theta_{ji}^a$, $j = 1, 2, i = 1, \ldots, 6$, with the projections (3.34), we get that

$$
\theta_{ji}(t) \ge \theta_{ji}^a,\quad \tilde{\theta}_{ji}(t)f_{ji}(t) \le 0,\quad j = 1, 2, i = 1, \ldots, 6. \tag{3.43}
$$

Considering this condition, (3.42) can be simplified as

$$\dot{W} \leq -K_V \tilde{V}^2 - K_\mu \tilde{\mu}^2 - K_\beta \tilde{\beta}^2 - K_\alpha \tilde{\alpha}^2 - K_p \tilde{p}^2 - K_q \tilde{q}^2 - K_r \tilde{r}^2 - e_1^T K e_1 + e_1^T \delta$$
$$\leq -K_V \tilde{V}^2 - K_\mu \tilde{\mu}^2 - K_\beta \tilde{\beta}^2 - K_\alpha \tilde{\alpha}^2 - K_p \tilde{p}^2 - K_q \tilde{q}^2 - K_r \tilde{r}^2$$
$$-\frac{min\{K_\gamma, K_\chi\}}{2}||e_1||_2^2 + \frac{1}{2\,min\{K_\gamma, K_\chi\}}\delta_0^2, \tag{3.44}$$

where $e_1(t) = [\tilde{\gamma}, \tilde{\chi}]^T$, $K = diag\{K_\gamma, K_\chi\}$, and $\delta = [\delta_1, \delta_2]^T$ such that $||\delta|| < \delta_0$ for a known upper bound δ_0.

Integrating both sides over the time interval $[t, t+T]$, we have

$$\int_t^{t+T} \dot{W} d\tau \leq -\sum_{i=1}^{7} K_i \int_t^{t+T} \tilde{z}_i^2 d\tau - \frac{min\{K_\gamma, K_\chi\}}{2} \int_t^{t+T} ||e_1||_2^2 d\tau$$
$$+\frac{\delta_0^2}{2\,min\{K_\gamma, K_\chi\}} T, \tag{3.45}$$

where $\tilde{z}_i = \tilde{V}, \tilde{\mu}, \tilde{\beta}, \tilde{\alpha}, \tilde{p}, \tilde{q}, \tilde{r}$ for $i = V, \mu, \beta, \alpha, p, q, r$. Therefore, we have

$$\sum_{i=1}^{7} K_i \int_t^{t+T} \tilde{z}_i^2 d\tau + \frac{min\{K_\gamma, K_\chi\}}{2} \int_t^{t+T} ||e_1||_2^2 d\tau \leq F(t) + \frac{\delta_0^2}{2\,min\{K_\gamma, K_\chi\}} T, \tag{3.46}$$

from which we can conclude that all error signals $\tilde{x}$, that is, $\tilde{z}_i$ and $e_1(t) = [\tilde{\gamma}, \tilde{\chi}]^T$ in the closed-loop control system are bounded and can be constrained in a small region by appropriate choice of gains like K_γ, K_χ. The uncertain parameters θ_{1i}, θ_{2i} are also uniformly bounded.

3.4 Results and Discussions

The adaptation gains $\gamma_{1i} = 1, \gamma_{2i} = 100000, \forall i = 1, \ldots, 6$ are chosen. The choice of controller gains in applications with parametric uncertainties is critical [17, 18]. The parameters $\theta_{1i}^*, \theta_{2i}^*$ are decided based upon experiments using wind tunnel testing [19]. The constant controller gains chosen as per the inequalities (3.23) indicate the necessary bandwidths, as presented in Table 3.1 [20].

The approximation errors can be reduced by increasing K_γ and K_χ without violation of the bandwidth limits. The mass of the tailless aircraft and the acceleration due to gravity are given by $m = 419.3$ slugs, $g = 32.17$ ft/s^2 respectively. The product inertias of the aircraft are also given above. Using these product inertias, we obtain the parameters $c_i, i = 1, \ldots, 7$, and $a_{ij}, i, j = 1, \ldots, 6$, are

$$a_{11} = a_{12} = 22.92,\ a_{13} = 3.06,\ a_{21} = 5.99,\ a_{22} = 8.04,\ a_{23} = -2.56,$$
$$a_{31} = 0.84,\ a_{32} = 5.35,\ a_{33} = -11.77,\ a_{14} = -a_{11},\ a_{15} = a_{12},\ a_{16} = a_{13}$$
$$a_{24} = a_{21},\ a_{25} = a_{22},\ a_{26} = a_{23},\ a_{34} = a_{31},\ a_{35} = a_{32},\ a_{36} = -a_{33}, \tag{3.47}$$

Table 3.1 Controller gains and Product inertias

Controller gains	K_V	K_γ	K_χ	K_μ	K_α	K_β	K_p	K_q	K_r
	1	0.8	0.8	4	4	2	20	20	10
Product inertias	I_{xx}	I_{yy}	I_{zz}	I_{yx}	I_{yz}	I_{xz}			
	11000	6500	18500	0	0	0			

such that from (3.30), we get

$$A = \begin{bmatrix} 0 & 45.84 & 6.12 \\ 11.98 & 16.08 & -5.12 \\ 1.68 & 10.70 & 0 \end{bmatrix},$$

and the control laws $u_{di}(t), i = 1, 2, 3$, and from (3.47), we get $u_{d4}(t) = -u_{d1}(t)$, $u_{d5}(t) = u_{d2}(t)$, $u_{d6}(t) = -u_{d3}(t)$.

The preferred trajectory is governed by these reference states:

$$\begin{aligned} \gamma_r &= \gamma(t_0) + 0.5\, sin(0.5t) \text{ deg}, \ \ \chi_r = \chi(t_0) + 0.5\, sin(0.5t) \text{ deg}, \\ V_r &= V(t_0) + 1.5\, sin(0.25t) \text{ ft/s}, \end{aligned} \tag{3.48}$$

where the initial flight bath angles are $\gamma(t_0) = 12.75$ deg and $\chi(t_0) = 0$ deg, and the initial velocity is $V(t_0) = 500$ ft/s. The initial body axis rates are $p(t_0) = q(t_0) = r(t_0) = 0$. The flight path tracking performance using a fixed nonadaptive inverse controller is shown in Fig. 3.3. The γ tracking error is not within acceptable bounds.

Figure 3.4 shows significantly improved tracking using AIC, especially for γ tracking error, although the velocity error term is not affected any differently to that which is observed in Fig. 3.3 by the usage of adaptive control because there is no parametric adaptive update in (3.10).

The corresponding nominal and adaptively estimated parameters are described in Figs. 3.5, 3.6 and 3.7.

It is clear from these plots that the parameter errors stay bounded, even though such errors do not go to zero unlike the tracking errors shown in Fig. 3.4.

The voltage signals applied to the six arrays are shown in Figs. 3.8, 3.9 and 3.10. The tracking errors for the mechanical control surfaces are not significantly different to the tracking errors. Note that the mechanical tracking is equivalent to the deployment of SJs with an ideal, fixed inverse control design with nominal parameters. We assume that the mechanical surfaces simply produce the commanded lift.

Next, we study overall performance under measurement error. We consider that the primary sensor aboard the aircraft is the IMU (inertial measurement unit, which is usually high quality and not very noisy). The IMU provides linear and rotational accelerations. Additive white gaussian noise with a standard deviation of 0.07 deg/s is added to the rotational velocities p, q, r. We observe changes, but the performance

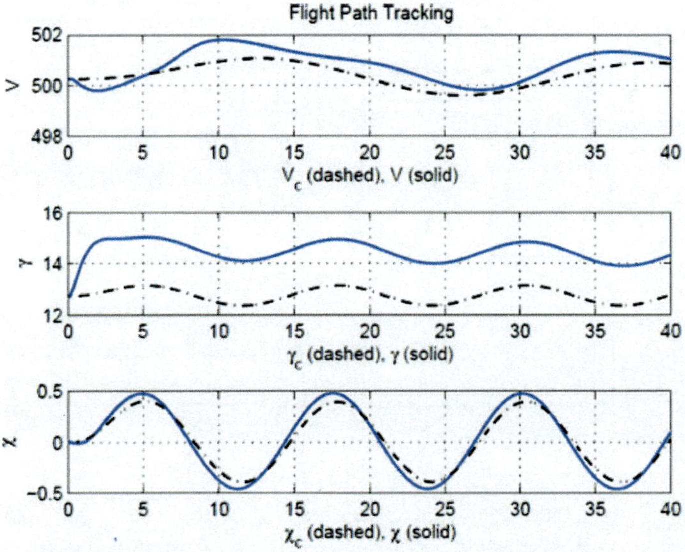

Fig. 3.3 Tracking errors for nonadaptive nonlinearity inverse controller

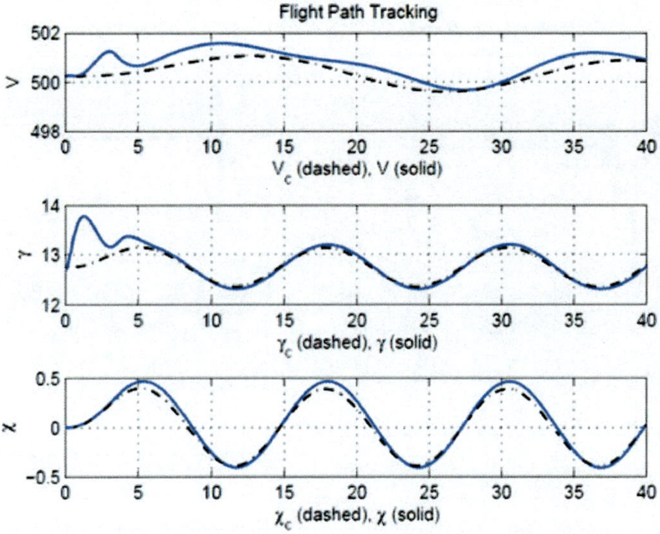

Fig. 3.4 Tracking errors for adaptive nonlinearity inverse controller

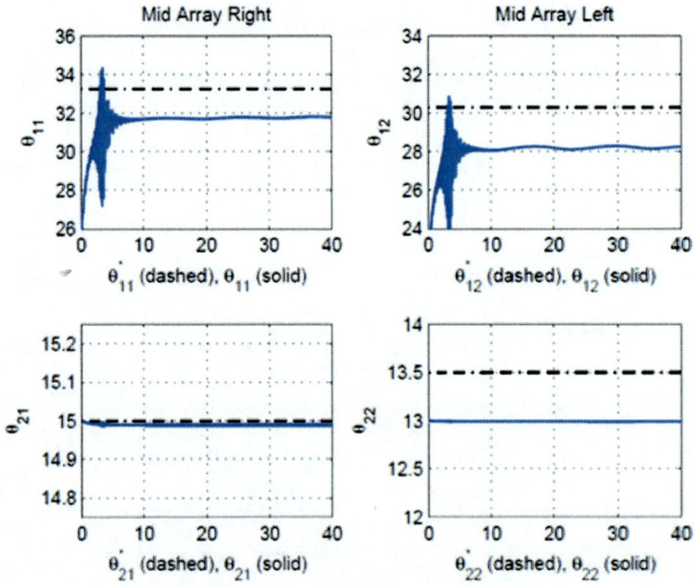

Fig. 3.5 True and estimated parameters for Middle flap arrays

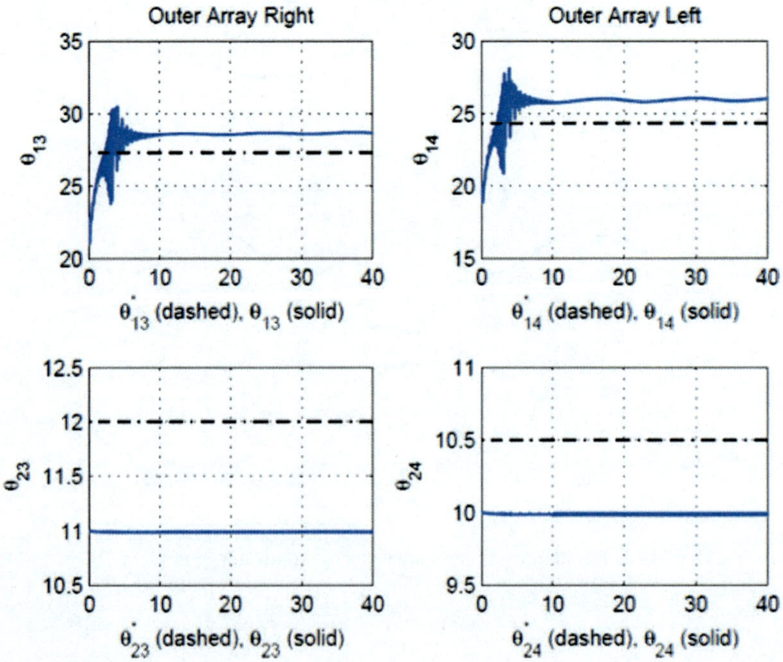

Fig. 3.6 True and estimated parameters for Outer flap arrays

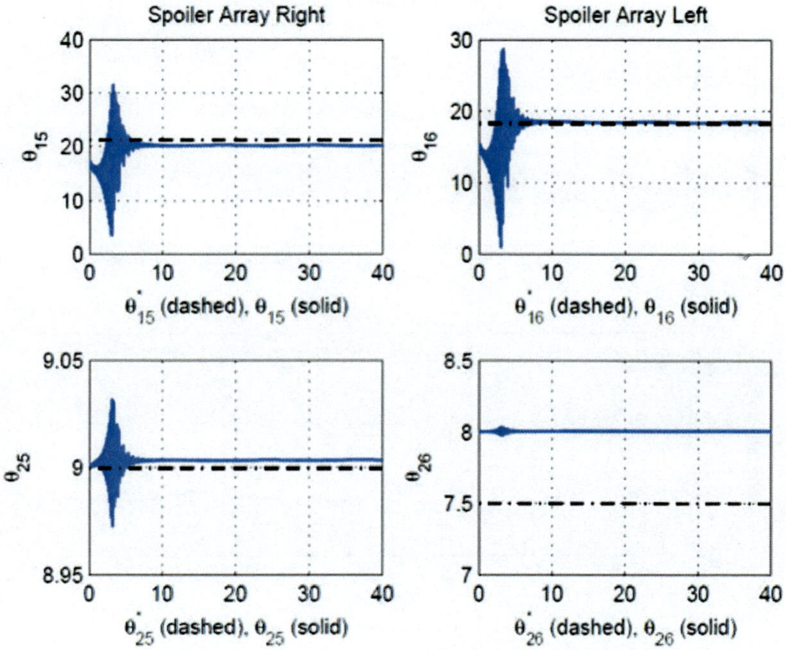

Fig. 3.7 True and estimated parameters for Spoiler arrays

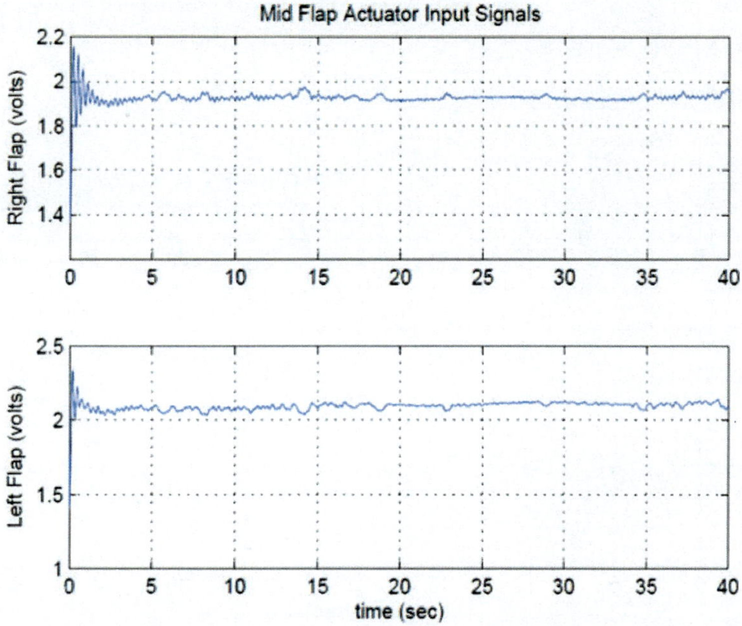

Fig. 3.8 Voltage signals for the mid-flap arrays

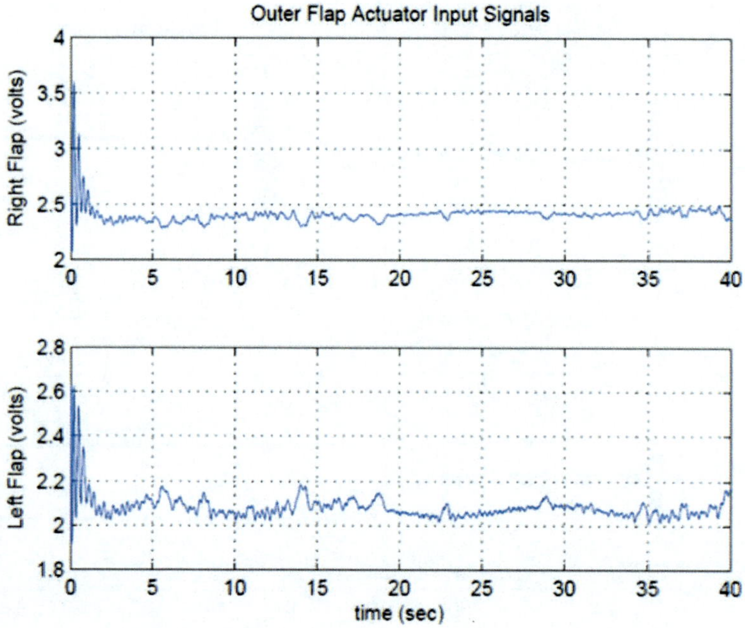

Fig. 3.9 Voltage signals for the outer-flap arrays

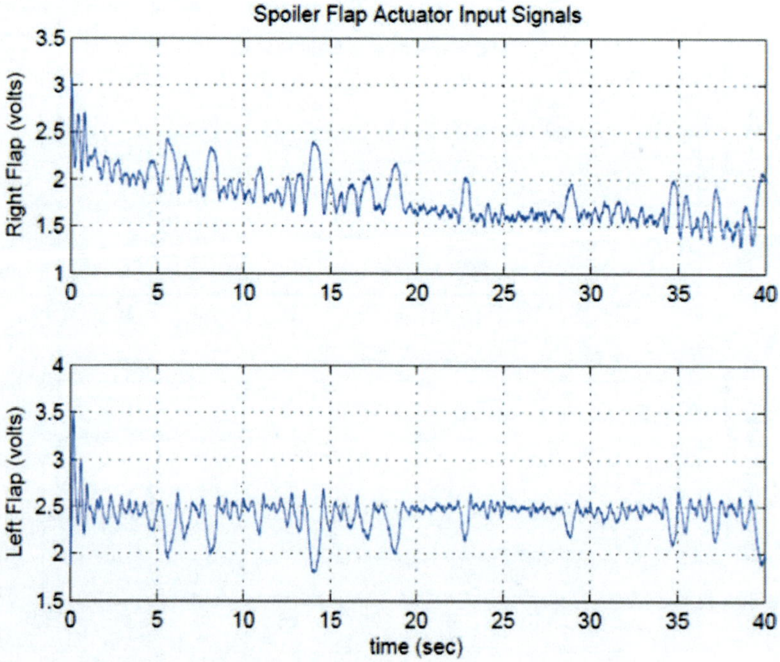

Fig. 3.10 Voltage signals for the spoiler arrays

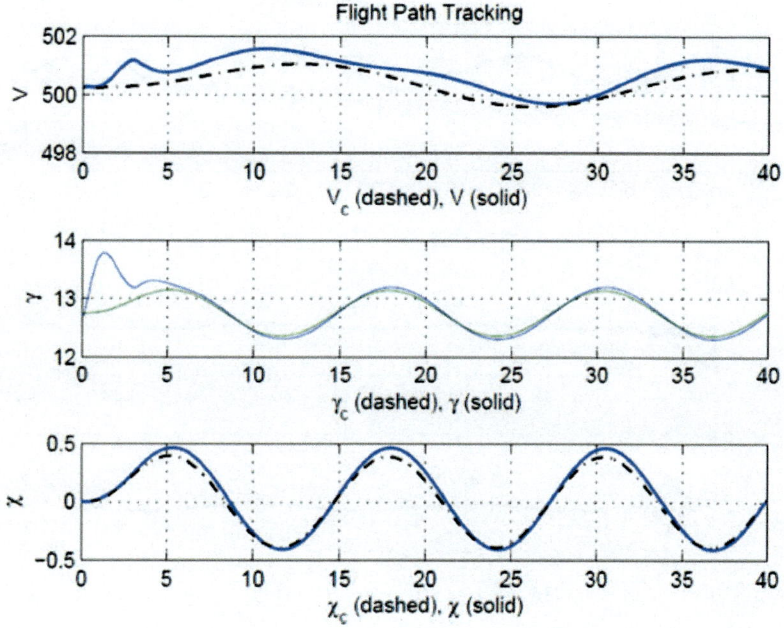

Fig. 3.11 Tracking errors for adaptive nonlinearity inverse controller in the presence of measurement error

is still reasonably good. The tracking errors are almost unchanged in Fig. 3.11. The tracking errors for the states p, q, r are shown in Fig. 3.12 and are reasonably good after some transience.

The parameter errors are shown in Figs. 3.13, 3.14 and 3.15 to have minimal deterioration when exposed to measurement errors.

We present an adaptive compensation control scheme for virtual aerodynamic wing shaping with arrays of SJAs. The effectiveness of the proposed adaptive control scheme applied to a nonlinear aircraft model is verified through simulations and it is expected that with adequately distributed SJAs, we would accomplish better performance compared to the usage of physically deflecting surfaces for control.

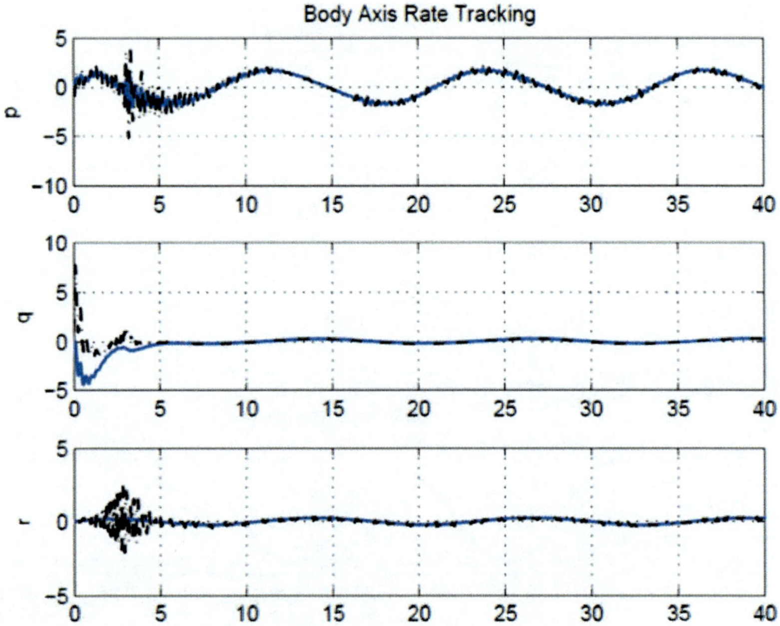

Fig. 3.12 Body axis tracking errors in the presence of measurement error

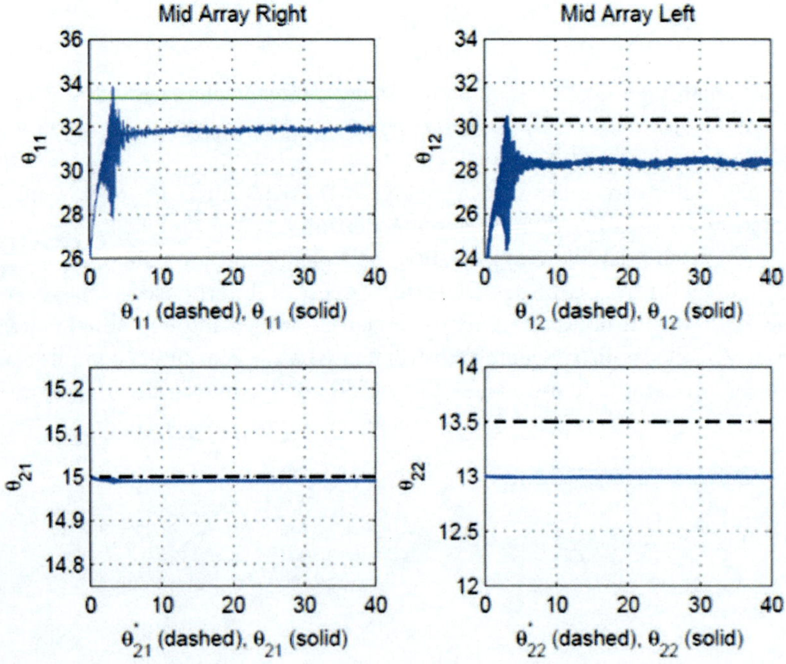

Fig. 3.13 Middle arrays (true and estimated parameters) with measurement error

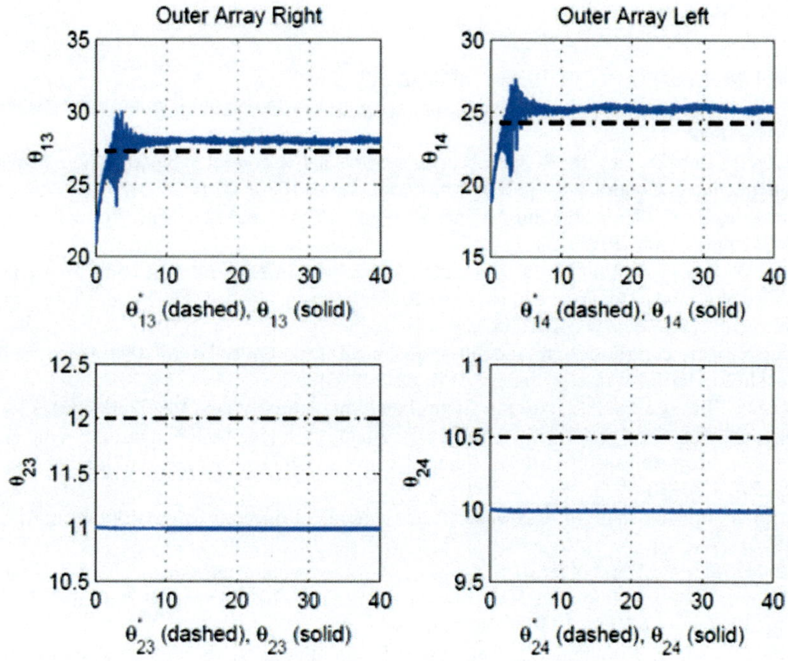

Fig. 3.14 Outer arrays (true and estimated parameters) with measurement error

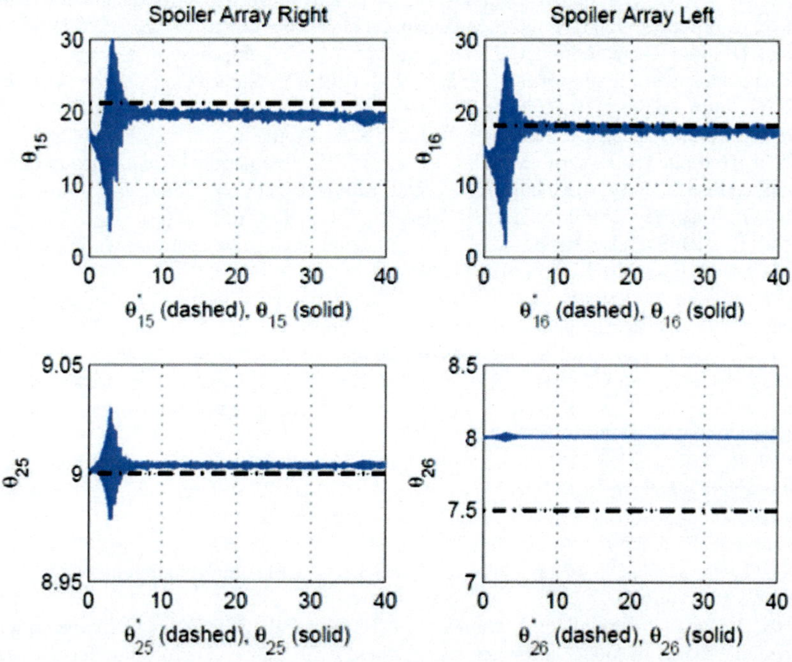

Fig. 3.15 Spoiler arrays (true and estimated parameters) with measurement error

References

1. Patel, R., Deb, D., Dey, R., Balas, V.E.: Adaptive control of single chamber two-population MFC. Intelligent Systems Reference Library, pp. 81–89 (2019). https://doi.org/10.1007/978-3-030-18068-3_7
2. Nath, A., Deb, D., Dey, R.: Robust observer based adaptive control of blood glucose in diabetic patients. Int. J. Control 1–14 (2020). https://doi.org/10.1080/00207179.2020.1750705
3. Franklin, G., Powell, J., Emami-Naeini, A.: Feedback Control of Dynamic Systems. Prentice-Hall, Upper Saddle River (2002)
4. Deb, D., Tao, G., Burkholder, J., Smith, D.: An adaptive inverse control scheme for a synthetic jet actuator model. In: Proceedings of the 2005, American Control Conference (2005). https://doi.org/10.1109/acc.2005.1470367
5. Multivariable control design guidelines: Technical Report WL-TR-96-3099. Honeywell and Lockheed Martin, Wright Patterson AFB, OH (1996)
6. Stevens, B.L., Lewis, F.L.: Aircraft Control and Simulation. Wiley, New York (1992)
7. Deb, D., Tao, G., Burkholder, J., Smith, D.: Adaptive compensation control of synthetic jet actuator arrays for airfoil virtual shaping. J. Aircraft **44**(2), 616–626 (2007). https://doi.org/10.2514/1.24910
8. Bryson, A.E., Jr.: Control of Spacecraft and Aircraft. Princeton University Press, Princeton (1994)
9. Deb, D., Tao, G., Burkholder, J., Smith, D.: Adaptive synthetic jet actuator compensation for a nonlinear tailless aircraft model at low angles of attack. In: 2006 American Control Conference. https://doi.org/10.1109/acc.2006.1655376
10. Deb, D., Tao, G., Burkholder, J.: High-order design of adaptive inverses for signal-dependent actuator nonlinearities. In: 2008 American Control Conference (2008). https://doi.org/10.1109/acc.2008.4587029
11. Deb, D., Tao, G., Burkholder, J., Smith, D.: An adaptive inverse control scheme for synthetic jet actuator arrays (2005). Infotech@Aerospace. https://doi.org/10.2514/6.2005-7170
12. Deb, D., Tao, G., Burkholder, J.: An adaptive inverse compensation scheme for signal-dependent actuator nonlinearities. In: 2007 46th IEEE Conference On Decision And Control (2007). https://doi.org/10.1109/cdc.2007.4434743
13. Deb, D., Sonowal, S.: Synthetic jet actuator based adaptive neural network control of nonlinear fixed pitch wind turbine blades. In: 2013 IEEE International Conference On Control Applications (CCA) (2013). https://doi.org/10.1109/cca.2013.6662759
14. Tao, G.: Adaptive Control Design and Analysis. Wiley, New York (2003)
15. Patel, R., Deb, D., Dey, R., Balas, V.E.: Introduction to adaptive control. Intelligent Systems Reference Library, pp. 53–65 (2019). https://doi.org/10.1007/978-3-030-18068-3_5
16. Nath, A., Deb, D., Dey, R., Das, S.: Blood glucose regulation in type 1 diabetic patients: an adaptive parametric compensation control-based approach. IET Syst. Biol. **12**(5), 219–225 (2018). https://doi.org/10.1049/iet-syb.2017.0093
17. Liu, Y., Deb, D., Tao, G.: Modeling and multivariable adaptive control of aircraft with synthetic jet actuators. In: 2008 7th World Congress On Intelligent Control And Automation (2008). https://doi.org/10.1109/wcica.2008.4593263
18. Patel, R., Deb, D., Dey, R., Balas, V.E.: Model reference adaptive control of microbial fuel cells. Intelligent Systems Reference Library, pp. 109–121 (2019). https://doi.org/10.1007/978-3-030-18068-3_10
19. Chatlynne, E., Rumigny, N., Amitay, M., Glezer, A.: Virtual aero-shaping of a Clark-Y airfoil using synthetic jet actuators. In: 39th Aerospace Sciences Meeting And Exhibit (2001). https://doi.org/10.2514/6.2001-732
20. Deb, D., Tao, G., Burkholder, J., Smith, D.: Adaptive synthetic jet actuator compensation for a nonlinear aircraft model at low angles of attack. IEEE Trans. Control Syst. Technol. **16**(5), 983–995 (2008). https://doi.org/10.1109/tcst.2007.912124

Chapter 4
Adaptive Compensation at High Angles of Attack

High AoA flight poses unique aerodynamic challenges, and these issues are studied extensively over the years [1]. An air vehicle at high AoA allows the pilot to rapidly nose-point to ensures a decisive action on the enemy. That is why next-generation aircraft must be able to operate consistently at higher and higher AoA than before and in a controlled fashion, thereby posing newer challenges in control, especially with the possible deployment of SJAs. Freezing conditions can also promote a local high AoA flow condition where ice accumulation on the tail leading edge may have an equivalent effect of a full-span stall strip [2]. The trend in the next-generation air vehicle design is towards increased exploitation of high-AoA aerodynamics and flow-field interactions. The relevance of high AoA studies is widespread, contrary to the general understanding that the pilot of a light single-engine propeller-driven air vehicle may also encounter high AoA aerodynamics as frequently as the pilot of a modern high-performance fighter.

Boundary layer flow separation is an undesirable phenomenon in almost all technical scenarios, and especially the point of flow separation and the AoA at which such an event occurs on the wings of air vehicles is of interest in this work. Flow separation causes loss of lift and (or) increased drag and reduction in retained energy. Therefore, there has been extensive research on the study of flow separation and ways and means to eliminate or delay the event. At high AoA, there is diminished flow attachment, and the separation takes place further and further away from the leading edge, thereby providing a diminished lift force. The resultant flows get controlled by airframe shaping or actively ensured by mechanical, pneumatic or active flow control devices like SJAs. In recent work, Linehan et al. demonstrate a servo-controlled sliding alula actuator to reorient separated flows at high AoA without undesired transients caused by the shedding effect of a stall vortex using a customized autopilot board and a 9-axis IMU for feedback control [3]. Other actuators which are used in high AoA separation control include weakly ionized plasma actuators consisting of

D. Deb et al., *Adaptive Compensation of Nonlinear Actuators for Flight Control Applications*, Studies in Systems, Decision and Control 386,
https://doi.org/10.1007/978-981-16-4161-9_4

two electrodes with a dielectric material in between [4]. Atkinson et al. [5] present a numerical investigation of such actuators to evaluate flow control over reentry vehicles with magnetically accelerated surface discharges at high AoA. In a recently published work, Zheng et al. have studied sliding discharge plasma actuators at high AoA operation of a slender body model [6]. Spark discharge thermodynamically drives a hybrid form of the SJAs and the plasma actuators to create high frequency (>5 kHz) and high-velocity (>300 m/s) pulsed jets [7].

As seen so far in this book, AoA have a direct bearing on the flow fields around the aircraft wing. We have seen in the earlier chapters that low AoA airfoil orientation is suitable for virtual shaping. However, at higher AoA, the characteristics of SJAs are very different, as will be seen in this chapter, and so the AIC technique also needs to be in line with this change. In this chapter, we explore adaptive compensation schemes to ensure aerodynamic flow control of an airfoil embedded with SJAs and operating at high AoA using wind tunnel tests and appropriate model approximations. High AoA aerodynamics is heavily dependent on wind tunnel testing. We present wind tunnel test results obtained using SJs on the leading edge of a wing based on NASA Wing 9 Planform at high AoA (16° – 24°) for flow separation delay. These test results provide nonlinearly parametrized actuator array models at an angle of attack of 24°. With the usage of two different parametrizations, the AIC method cancels the unknown actuator nonlinearities. There is no need to have an exact knowledge of the parametric values, and only bounds of such parameters are sufficient information.

4.1 Synthetic Jet Actuator Model Development

At the outset, it is important to have some basic understanding about wind tunnels. Wind tunnels are large tubes equipped with various sensors, with controlled stream of air moving inside for replication of actions of an object in flight by using small and properly scaled static versions of the same. The wind tunnels make it seem like the object is indeed flying. Wind tunnels are useful in verifying aerodynamic theories and for facilitating different aspects of aircraft design. Apart from aircraft and spacecraft, aerodynamic studies in wind tunnels are profitable for the solution of solving design problems in various other disciplines like automobiles, trains, and buildings, boats, and bridges. Wind tunnels effectively obtain the complete lift and drag on a body and the surface distributions at specific points. We envisage that instead of being substituted by numerical simulations, wind tunnels are expected to be complementary in nature.

In this section, we describe the results obtained through a model test planform in a low-speed wind tunnel facility at the U.S. Air Force Academy (USAFA). We examine specific implementation challenges, to acquire experimental measurements for control development, and to develop nominal actuator models. The wing model used in this study is based on the NASA Wing 9 platform [8] and is a prototype of a flying Delta Wing UAV with an isosceles triangular shaped lifting body. The leading-edge and trailing-edge sweep angles are 50°. The wingspan is 0.608 m, and

Fig. 4.1 SJA integrated tailless aircraft model at the leading edge in the wind tunnel

the mid-span chord is 0.483 m, providing a planform area of 0.145 m^2. Active flow control takes place with eight actuators integrated into the leading edge of the wing under test to aid in flow reattachment when the flow separates away from the leading edge of the surface at high AoA. Two piezoelectric diaphragms drive an actuator at a frequency of 1940 Hz. The orifice dimensions are 0.5 mm × 32 mm. The pitch angle is varied from 16° to 26° at zero yaw and roll angles. A sting balance measures the aerodynamic forces and moments.

The interior view of the wind tunnel which houses an actual Tailless aircraft model with SJA integrated at the leading edge in the wind tunnel, is shown in Fig. 4.1.

Figures 4.2a, b show the top and side views of the wing, indicating the SJAs, pressure ports, and the access ports. The SJs are the rectangular shapes that run along the leading edge because it is the best place to perform wing control at high AoA close to the stall. A critical aspect of stall is an abrupt aerodynamic performance change of the associates devices that may result in a sudden lift and drag alteration due to separated flow.

Without loss of generality, there are eight actuators in all in this test setup, with the same cavity size and orifice dimensions. Each actuator is individually addressable. The small circles (viewed as dots here) along the surface running from the leading to the trailing edge are the surface pressure ports. Pressure ports are small holes drilled into the surface that is connected to a high sensitive pressure transducer and provide a measurement of the pressure distribution over the surface of the model.

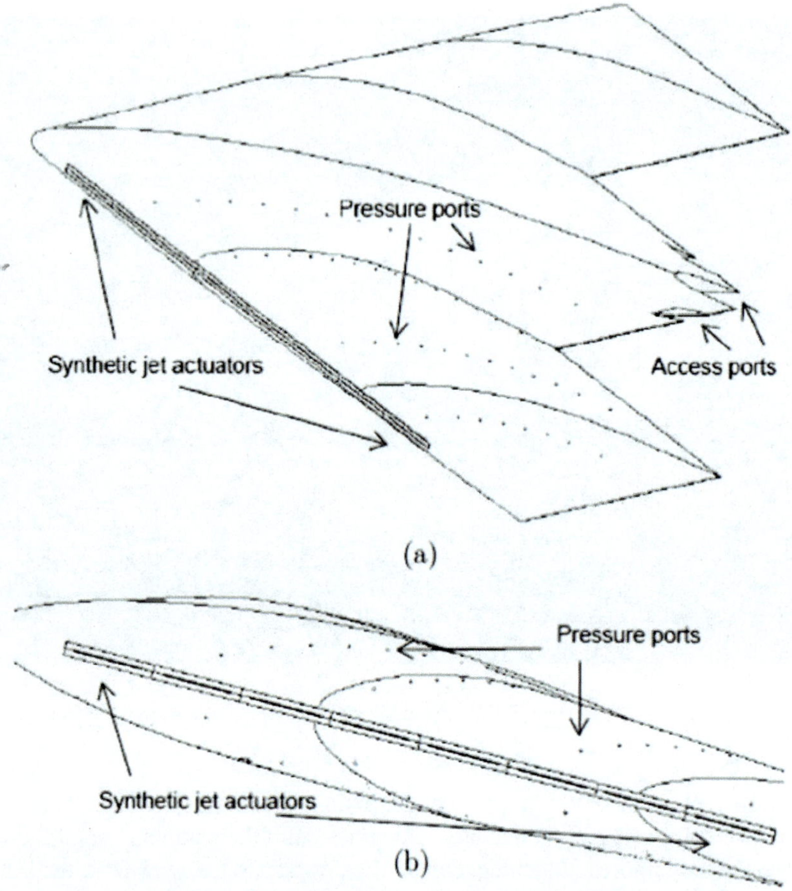

Fig. 4.2 Wing model with SJAs along the leading edge: **a** top view, **b** side view

Measurements of air pressure yield information about the total wind load distribution. Different types of dynamic pressure measurements are possible such as turbulence, noise, sound, cavitation and pulsations. The cut-outs at the mid-span trailing edge are access ports for the mounting sting and wiring to the actuators.

In the current study, the control effect is separation control with SJ array along the leading edge at post-stall AoA. Figure 4.3 shows lift coefficient C_L variations for the baseline and the controlled configurations with all eight of the actuators in the array activated, for three different values of C_μ, at Reynolds number (root-chord) of $Re_c = 5.30 \times 10^5$ and free-stream velocity of 23 m/s. For this configuration, there is no apparent dependence of the control effect on the value of C_μ, indicating that beyond the normal stall angle the impact of SJAs is practically uniform over a range of momentum coefficient values.

Figure 4.4 show variations of the pitch moment coefficient C_M for the baseline and the controlled configurations with all eight of the actuators in the array activated,

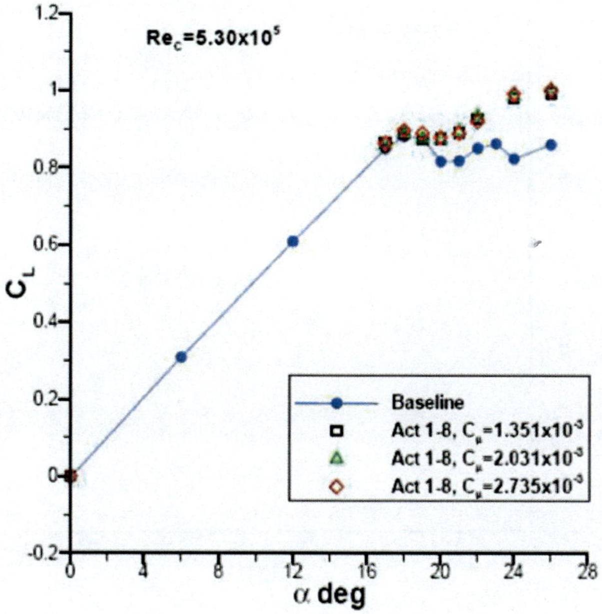

Fig. 4.3 Variations of lift coefficient with angle of attack

and with the same values of Reynolds number, free-stream velocity and C_μ as in Figure 4.3. For this control configuration, there is no apparent dependence on the control effect on the value of C_μ. Note that the SJAs are placed only along one side of the wing, and the changes in the aerodynamic coefficients are representative of one-half of the effect expected should the entire span of the wing contain actuators.

We also examine the control effect by operating six actuator subsets and compare that to the case when all of the actuators are activated. Three of these subsets focus on the impact of the in-board actuators, and the three others focus on the effect of the out-board actuators. Three subsets of in-board actuators are the 1–4, 1–5, and 1–6 sets where the actuator number increases with distance from mid-span. The effect of the in-board actuators is visible in Fig. 4.5.

For AoA at (19°) and just beyond (22°) stall, the control effect for all three subsets is the same with only a modest improvement in C_L. At $\alpha = 24°$, the control effect is one half of that which results when all of the actuators activated. The same phenomena observed for all three subsets of actuators. Only at $\alpha = 26°$ is there a proportional effect arising from the number of actuators activated with the 1-to-4 subset, proving only half as efficient as the 1-to-5 and 1-to-6 subsets. Note that as the reported C_μ is the sum of the C_μ values for the individual actuators, the effective C_μ is different for the different subsets and decreases as the number of actuators activated decreases. That the different subsets have equal effectiveness at AoA up to 26° indicates that beyond a critical value of C_μ, there is limited dependence of the control effect on C_μ.

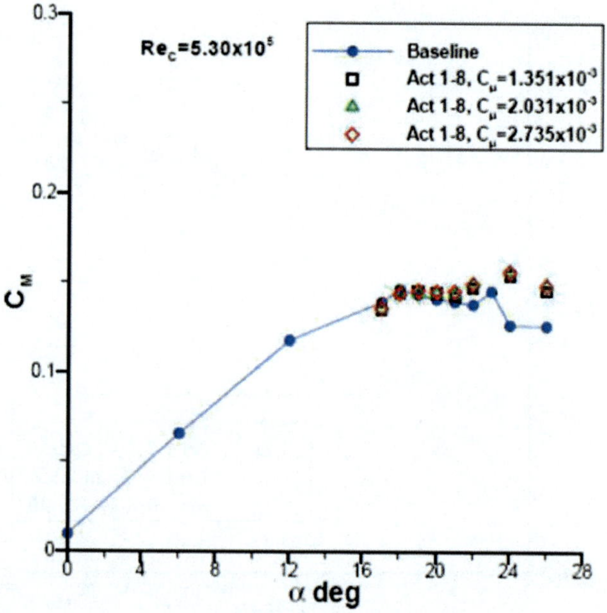

Fig. 4.4 Variations of pitch moment coefficient with angle of attack

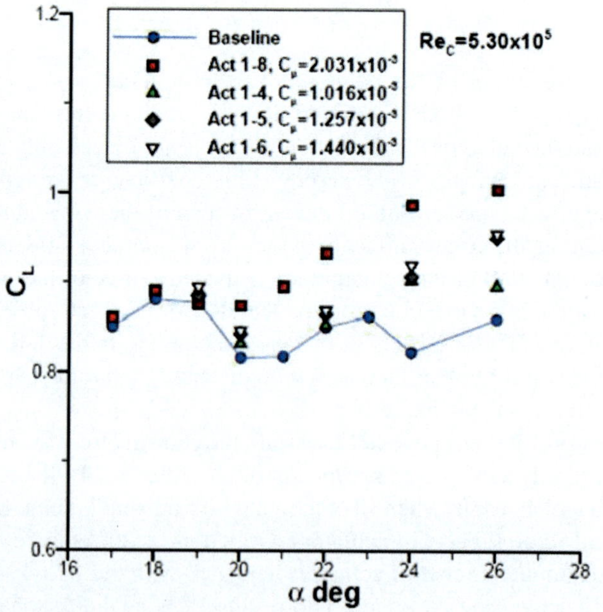

Fig. 4.5 Variations of lift coefficient for baseline uncontrolled case and the controlled cases with in-board subsets of actuators activated

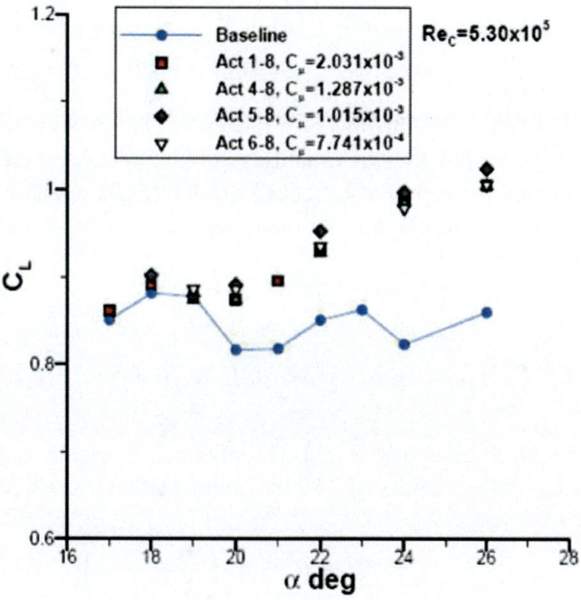

Fig. 4.6 Variations of lift coefficient for baseline uncontrolled case and the controlled cases with out-board subsets of actuators activated

The effect of the outboard actuators is in Fig. 4.6. The three subsets of outboard actuators are the 4-to-8, 5-to-8, and 6-to-8 sets of actuators. At all AoA and for the individual actuator C_μ value used, there is no apparent effect on the control of changing the number of actuators activated. That is, the three outer-most actuators, 6-to-8, are as effective at controlling the separation as all eight actuators. As the separation begins near the wingtips and progresses in-board with the angle of attack, locating the control effect along the wing where separation occurs is as effective as applying control along the entire semi-span.

Figures 4.3, 4.5, and 4.6 also show improvement in lift and moment coefficients over the baseline model, beyond stall, thereby validating the control effect produced by activation of jets on the leading edge for post-stall separation control. This result emphasizes that the strategic placement of the control is as important as the amount of control input to the flow.

4.2 Nonlinearity Model of Synthetic Jet Actuators

At high AoA, the nonlinear characteristics of the SJs are significantly different from that for low AoA. In the data reported by Kondor et al. [9], there is a clear, but limited, influence from the momentum coefficient C_μ on the lift force at high AoA. The reason for this observation is that the control action at the leading edge is almost exclusively useful for separation control at high AoA. The wind tunnel tests performed at USAFA by the researchers at the University of Wyoming validate these results, and the implications of SJ control at a wider range of high AoA are studied, as already described in detail in Figs. 4.3 and 4.5. At AoA ($\alpha = 16° - 24°$), where the SJs are controlling the flow separation, there is very little, or no, change in C_L with variations in C_μ, because control at the leading edge does not change, in any significant way, the flow over the wing. The variation of the pitch moment coefficient C_M is negligible with variations in C_μ. These coefficients of lift, drag, and momentum are the combined effect of the active SJAs. There are other relevant studies on the force and pitching moment variations at high AoA, for instance, as applied to race cars [10].

4.2.1 Nonlinearity Models with Partially Known Parameters

Study of the experimental data obtained from the wind tunnel tests reveals that at AoA in the vicinity of 24°, the control effect given by C_L and C_M are

$$C_L = \theta_2^{h*} + \theta_1^{h*} \sin^2\left(\frac{\pi}{2}\frac{C_\mu}{C_{\mu,max}}\right), \quad C_M = \theta_5^{h*} + \theta_4^h C_L, \tag{4.1}$$

for some nominal parameters $\theta_1^{h*}, \theta_2^{h*} > 0$ and $\theta_5^{h*} < 0$ where θ_1^{h*} represents the maximum range of the lift coefficient produced by the combined effect of the SJs which are active, θ_2^{h*} represents the minimum lift coefficient, and the parameters θ_5^{h*} and θ_4^h nominally determine the relationship of C_L with C_M. The parameter θ_4^h is assumed to be known and constant, as the test results show negligible variation.

Additionally, $C_{\mu,max}$ is the maximum momentum coefficient and is known from the dimensions of the jets ($C_{\mu,max} = 3 \times 10^{-3}$ in this study) and the momentum coefficient C_μ as defined in [11] is $C_\mu = \frac{pA_{pp}^2}{u_\infty^2}$, where u_∞ is the freestream velocity in the jets, p is an unknown constant governed by characteristics like cavity dimensions, number of activated jets, and $v = A_{pp}^2$ is the square of the input peak-to-peak voltage amplitude. Under conditions in which we know the parameters u_∞ and p, we have

$$C_L = \theta_2^{h*} + \theta_1^{h*} \sin^2(av), \tag{4.2}$$

where $a = \frac{\pi}{2}\frac{p}{u_\infty^2 C_{\mu,max}}$ is known. For the compensation of the nonlinear uncertainties, we consider that the

(i) measurement of the output variable C_M of the SJAA is not feasible, thereby necessitating development of desired control signals C_{Md} by state feedback;
(ii) parameters of the nonlinearity models represented by (4.2), that is, θ_1^{h*}, θ_2^{h*} and θ_5^{h*} are unknown, but their bounds are known.

4.2.2 Nonlinearity Models with Unknown Parameters

Under rapid changes in the actuator states (activated or inactivated), the assumption of the knowledge of the parameter p no longer holds. The physical limitation of the input voltage v is governed by $C_\mu \leq C_{\mu,max}$, and the upper bound of v is

$$v \leq \frac{C_{\mu,max} U_\infty^2}{p} = \frac{\pi}{2\theta_3^{h*}}, \tag{4.3}$$

where $\theta_3^{h*} = \frac{\pi p}{2C_{\mu,max}U_\infty^2}$ is a true nonlinearity parameter of the SJAA which is to be adaptively compensated in addition to the parameters θ_1^{h*} and θ_2^{h*}.

Using (4.2) and (4.3), for the parametrized nonlinearity model $N(\cdot)$, the lift coefficient $C_L(\mathrm{t})$ is generated by the SJAs at high AoA and constant density of airstream along the spanwise orifice length, and is approximately given as

$$C_L = \theta_2^{h*} + \theta_1^{h*} \sin^2\left(\theta_3^{h*} v\right). \tag{4.4}$$

For a SJA at 24° angle of attack, the typical parameters are $\theta_1^{h*} = 0.1, \theta_2^{h*} = 0.9$ and $\theta_3^{h*} = \frac{\pi}{32}$, with $C_{\mu,max} = 3 \times 10^{-3}$, $p = 7.5$, $u_\infty = 200$ft/sec (which implies that $v \leq 16$ as $C_\mu \leq C_{\mu,max}$). Figure 4.7 shows the relationship plots of C_L and v.

Figure 4.7a shows the variations at $\theta_1^{h*} = 0.1, 0.2, 0.3$ and $\theta_3^{h*} = \frac{\pi}{32}$, and Fig. 4.7b show the variations at $\theta_3^{h*} = \frac{\pi}{24}, \frac{\pi}{28}, \frac{\pi}{32}$ for $\theta_1^{h*} = 0.1$ (with $\theta_1^{h*} = 0.9$). The SJ characteristics given by (4.4) and (4.1) are not valid for v beyond the bound given by (4.3). The parameters of the nonlinearity model θ_1^{h*}, θ_2^{h*}, and θ_3^{h*} are unknown, but the bounds are known.

When the actuator nonlinearities are not in a simple form that is easily invertible, there is a need to consider an approximation based approach. Farrell et al. have provided a backstepping-based multi-loop controller using adaptive function approximation to control UAVs [12]. In a different work, Liang et al. have utilized sliding mode and uncertain dead zone actuator functional approximation for position tracking application in a DC motor [13]. Hofinger has explored an adaptive greedy algorithm to approximate nonlinear functions such that the rate of convergence is independent of dimension of the uncertain parameter-space [14].

Figure 4.8a shows the variations of lift coefficient approximation error are within the desired bounds for the entire range of variation. Figure 4.8b shows that the variations of pitch moment approximation error are within the desired bounds stated above.

An approximate linear parametrization is based on the approximated form of a nonlinearly parametrized function. For example, we can utilize a least-squares curve-fitting technique to find the polynomial $\hat{C}_L$ so that the approximation error $|\Delta C_L(v)| = |C_L(v) - \hat{C}_L(v)| < 3 \times 10^{-4}$. The pitch moment approximation error is given by $|\Delta C_M(v)| = |C_M(v) - \hat{C}_M(v)| < 4.5 \times 10^{-4}$, over the range $v(t) \in (0, 16)$.

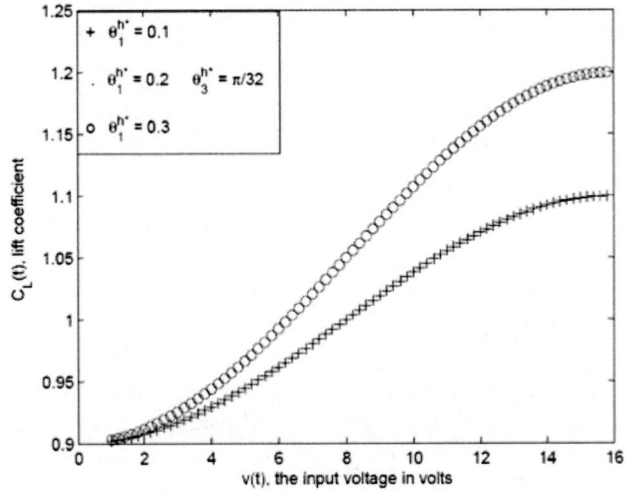

(a)

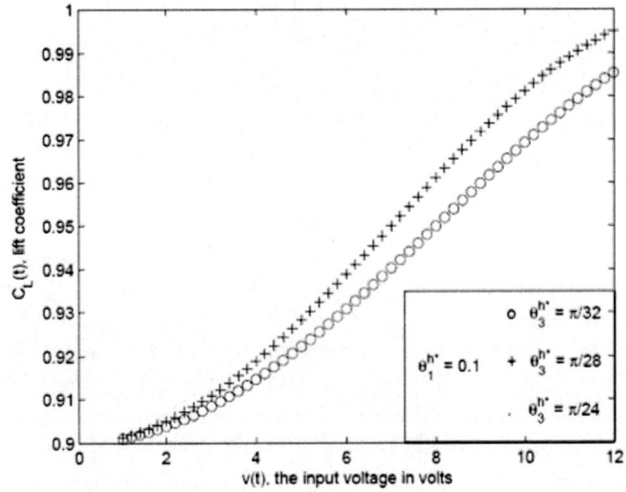

(b)

Fig. 4.7 Variations of lift coefficient versus actuator voltage magnitude: **a** different values of θ_1^{h*}, and fixed θ_2^{h*}, θ_3^{h*}, **b** different values of θ_3^{h*}, and fixed θ_1^{h*}, θ_2^{h*}

With $s(t) = v^2(t)$, such linearly parametrized approximated nonlinearity models are given by

$$\hat{C}_L(t) = \theta_1^* + \theta_2^* s(t) + \theta_3^* s^2(t) + \theta_4^* s^3(t), \quad \hat{C}_M(t) = \theta_5^{h*} + \theta_4^h \hat{C}_L(t), \tag{4.5}$$

for some nominal parameters $\theta_1^*, \ldots, \theta_4^*$, and θ_5^{h*} which can be estimated and which have the bounds $\theta_1^*, \theta_2^*, \theta_4^* > 0$, $\theta_3^* < 0$ and $\theta_5^{h*} < 0$.

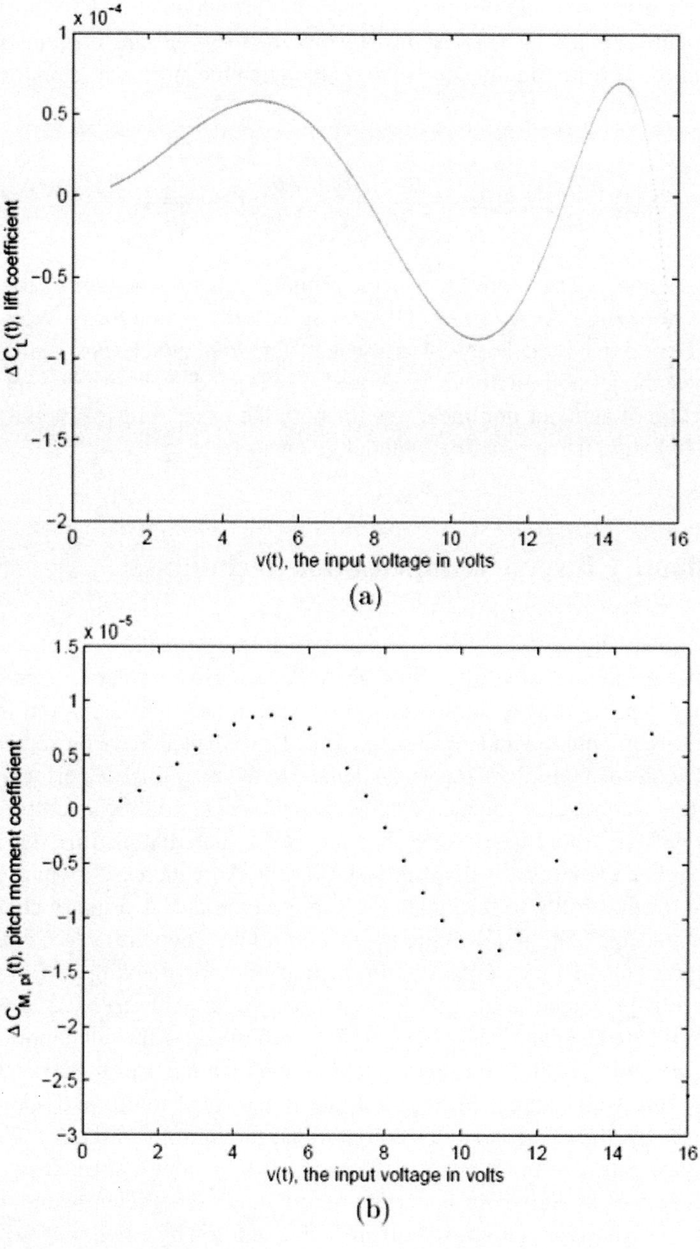

Fig. 4.8 Approximate coefficient errors against actuator voltage: **a** $\Delta C_L(t)$ versus $v(t)$, **b** $\Delta C_M(t)$ versus $v(t)$

Note that the approximation of C_M given by $\hat{C}_M$ is obtained from $\hat{C}_L$, and therefore the parameters θ_4^h and θ_5^{h*} appear in (4.5) as in (4.1) and are not independently approximated. The maximum value of $s(t)$, such that the approximation for actuator characteristics as given in (4.5) is valid, is given by

$$s_{max}(t) = -\frac{\theta_3^*}{3\theta_4^*} - \sqrt{\frac{1}{9}\left(\frac{\theta_3^*}{\theta_4^*}\right)^2 - \frac{\theta_2^*}{3\theta_4^*}}. \tag{4.6}$$

In this section, we have presented wind tunnel test results pertaining to SJ model development at high AoA (16° – 24°), using a Wing 9 planform. We have also developed a nominal model for SJA array at 24° angle of attack (post-stall). A cubic approximate model of the actuator array is also presented. Next, we present adaptive compensation of actuator nonlinearities for both the exact actuator model given by (4.1) and (4.2) and the approximate model given by (4.5).

4.3 Adaptive Inverse Compensation Techniques

Unlike the relatively simpler control techniques at low AoA [15–17], efficient feedback control techniques are critical in high AoA scenarios, primarily because of the sudden flow separation near stall, and control researchers have explored this problem in various air vehicles and different airfoils. Devaud et al. have presented several methods to control a missile in the vertical plane to increase missile performance and flight range and concluded that it is possible to use linear or nonlinear control methods to stabilize a highly nonlinear system like a highly manoeuvrable missile [18].

Control effectiveness typically diminishes as the AoA increases, which is particularly true for the ability to formulate the yawing moment. A primary challenge in the widespread application of SJAs for flight control has been the lack of closed-loop control algorithms that guarantee stability and asymptotic tracking performance. In the test results presented in the previous section, we have observed such nonlinear relationships between the input voltage signal amplitude and the pitch moment coefficient. Pedro and Mansfield present a detailed nonlinear aircraft model considering thrust vectoring as an augmented control signal along with a nonlinear dynamic inversion controller for pitch acceleration longitudinal mode of the aircraft at high AoA and post-stall considerations [19]. In another work, an optimal feedback method recovers a fighter aircraft with uncertain aerodynamic coefficients and stochastic parametric uncertainties at high AoA, from stall condition by employing polynomial chaos expansion [20].

An AIC framework for the SJA uncertainties $N(\cdot)$ is depicted in Fig. 4.9.

The system dynamics are represented by $\dot{x} = f(x) + g(x)C_M$, and $N(\cdot)$ is the SJA nonlinearities, $\widehat{NI}(\cdot)$ is the corresponding adaptive inverses to be developed, C_M is the control input, and C_{Md} represents the desired input for the SJA array. A linearized state feedback control law can then be generated.

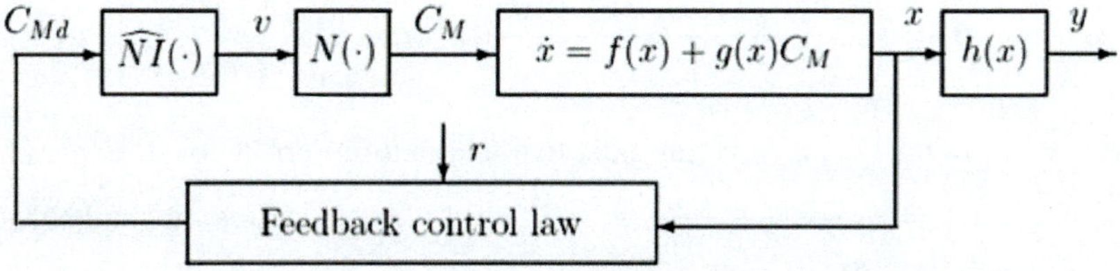

Fig. 4.9 Closed-loop inverse compensation system for pitch moment

4.3.1 Adaptive Inverse Compensation: Exact Actuator Model

We employ an appropriate adaptive inverse for the unknown nonlinearities given by

$$v(t) = \frac{1}{a}\arcsin\left(\sqrt{\frac{\theta_2^h(t) - C_{Ld}(t)}{\theta_1^h(t)}}\right), \tag{4.7}$$

where $\theta_1^h(t), \theta_2^h(t)$ are the parameter estimates and $C_{Ld}(t)$ is the desired lift coefficient, given by

$$C_{Ld}(t) = \frac{1}{\theta_4^h}\left(C_{Md}(t) - \theta_5^h(t)\right), \tag{4.8}$$

for desired state feedback control law $C_{Md}(t)$, adaptive estimate of θ_5^{h*}) given by $\theta_5^h(t)$, and the feedback control signal $C_{Md}(t)$ (to be generated).

The inverse parameters $\theta_i^h(t), i = 1, 2, 5$ are estimated using adaptive update laws and remain within known bounds (that is, $\theta_i^h > 0, i = 1, 2,$ and $\theta_5^h < 0$ because $\theta_i^{h*} > 0, i = 1, 2,$ and $\theta_5^{h*} < 0$, is ensured by adaptive update laws. We assume that $C_{Ld}(t) < \theta_2^h(t)$ such that the adaptive estimation function is implementable. As a consequence of this assumption and $\theta_1^h > 0$ we get $v(t) \neq 0$. Using (4.1) and (4.7), we obtain

$$C_L(t) = \frac{\theta_2^{h*}\theta_1^h(t) - \theta_1^{h*}\left(\theta_2^h(t) - C_{Ld}(t)\right)}{\theta_1^h(t)}. \tag{4.9}$$

Thus, the lift coefficient error is given by

$$C_L(t) - C_{Ld}(t) = \left(\frac{\theta_2^h(t) - C_{Ld}(t)}{\theta_1^h(t)}\right)\tilde{\theta}_1^h(t) - \tilde{\theta}_2^h(t), \tag{4.10}$$

where $\tilde{\theta}_1^h(t) = \theta_1^h(t) - \theta_1^{h*}$ and $\tilde{\theta}_2^h(t) = \theta_2^h(t) - \theta_2^{h*}$ are the adaptive estimation errors and where $\theta_i^h(t), i = 1, 2$ are updated. Additionally, using (4.1), (4.9), and (4.10), the moment coefficient error is given by

$$C_M(t) - C_{Md}(t) = \left(\frac{\theta_4^h \theta_2^h(t) - C_{Md}(t) + \theta_5^h(t)}{\theta_1^h(t)} \right) \tilde{\theta}_1^h(t) - \theta_4^h \tilde{\theta}_2^h(t) - \tilde{\theta}_5^h(t) = \tilde{\theta}^T(t)\omega(t), \tag{4.11}$$

where $\tilde{\theta}_5^h(t) = \theta_5^h(t) - \theta_5^{h*}$ is an adaptive estimation error for the parameter θ_5^{h*}, and $\omega(t) = \left[\frac{\theta_4^h \theta_2^h(t) - C_{Md}(t) + \theta_5^h(t)}{\theta_1^h(t)}, -\theta_4^h, -1 \right]^T$ is a regressor, and $\tilde{\theta}^h(t) = \left[\tilde{\theta}_1^h(t), \tilde{\theta}_4^h(t), \tilde{\theta}_5^h(t) \right]^T$ is the parameter error vector.

4.3.2 Adaptive Inverse Compensator: Approximate Model

From (4.5), we get the approximate actuator model

$$\hat{C}_M(t) = C_M(t) - \delta_M(t) = N_a(\theta_a^*; s(t)) = \theta_5^{h*} + \theta_4^h \left(\theta_1^* + \theta_2^* s(t) + \theta_3^* s^2(t) + \theta_4^* s^3(t) \right), \tag{4.12}$$

where $\hat{C}_M(t)$ is the approximated pitch moment coefficient, δM is the approximation error, and $\theta_a^* = \left[\theta_1^*, \theta_2^*, \theta_3^*, \theta_4^*, \theta_5^{h*} \right]^T$ is the parameter vector.

An adaptive inverse controller structure formulation with an approximate parametrized nonlinearity $N_a(\theta_a^*; s(t))$ requires a nonlinearity inverse $s(t) = v^2(t) = \widehat{NI}(C_{Md}(t))$ as a compensator. We generate the inverse signal $s(t)$ by solving the cubic equation

$$C_{Md}(t) = \theta_5^h(t) + \theta_4^h \left(\theta_1(t) + \theta_2(t)s(t) + \theta_3(t)s^2(t) + \theta_4(t)s^3(t) \right), \tag{4.13}$$

where $\theta_a(t) = \left[\theta_1(t), \theta_2(t), \theta_3(t), \theta_4(t), \theta_5^h(t) \right]^T$ are the adaptive estimates of θ_a^*. These estimates have the bounds: $\theta_1(t), \theta_2(t), \theta_4(t) > 0, \theta_3(t), \theta_5^h(t) < 0$.

Adaptive laws update and ensure through parameter projection that the inverse parameters $\theta_a(t)$ stay in prespecified regions. As explained earlier, (4.5) represents an approximated form $N_a(\cdot)$ of the actuator array nonlinearity model $N(\cdot)$, given by (4.2). The adaptive inverse for the approximated nonlinearity model $N_a(\cdot)$ is given as $\widehat{NI}_a(\cdot)$ and shown in (4.13). Therefore, this adaptive inverse model is effectively an approximation of the inverse model for the original SJA array nonlinearity model shown in (4.5).

Substituting $s(t) = p(t) - \frac{\theta_3(t)}{3\theta_4(t)}$ in (4.13), we get a cubic equation [21]

$$p^3(t) + \phi_1(t)p(t) + \phi_2(t) = 0, \quad p(t) > \frac{\theta_3(t)}{3\theta_4(t)}, \tag{4.14}$$

whose parameters $\phi_1(t), \phi_2(t)$ are given by

$$\phi_1(t) = \frac{\theta_2(t)}{\theta_4(t)} - \frac{\theta_3^2(t)}{3\theta_4^2(t)},$$

$$\phi_2(t) = \frac{2\theta_3^3(t)}{27\theta_4^3(t)} - \frac{\theta_2(t)\theta_3(t)}{3\theta_4^2(t)} + \frac{\theta_1(t) - \frac{1}{\theta_4^h}\left(C_{Md}(t) - \theta_5^h(t)\right)}{\theta_4(t)}. \quad (4.15)$$

As shown graphically by Hellman [22] from an analysis of cubic roots, the solutions to (4.14) depend upon the sign of $\frac{\phi_1^3(t)}{27} + \frac{\phi_2^2(t)}{4}$. Solutions of cubic equations expressed with the quadratic term suppressed are widely available widely in literature [23]. We can determine additional conditions such that the estimated inverse $s(t)$ is positive and satisfies the conditions of parametric bounds.

A control error $\hat{C}_M(t) - C_{Md}(t)$ expression is vital in the design of adaptive update laws. Using (4.12) and (4.13), we get

$$\hat{C}_M(t) - C_{Md}(t) = -\tilde{\theta}_5^h(t) - \theta_4^h\left(\tilde{\theta}_1(t) + \tilde{\theta}_2(t)s(t) + \tilde{\theta}_3(t)s^2(t) + \tilde{\theta}_4(t)s^3(t)\right), \quad (4.16)$$

where $\tilde{\theta}_i(t) = \theta_i(t) - \theta_i^*, i = 1, \ldots, 4,$ and $\tilde{\theta}_5^h(t) = \theta_5^h(t) - \theta_5^{h*}$ are the adaptive estimation errors, $s(t)$ is a function of the desired moment coefficient $C_{Md}(t)$. Using (4.13) and (4.16), the moment coefficient error is given by

$$C_M(t) - C_{Md}(t) = -\tilde{\theta}_a(t)\omega(t) + \delta_M(t), \quad (4.17)$$

where $\delta_M(t)$ is the approximation error, $\tilde{\theta}(t) = \left[\tilde{\theta}_1(t), \tilde{\theta}_2(t), \tilde{\theta}_3(t), \tilde{\theta}_4(t), \tilde{\theta}_5^h(t)\right]^T$ is the parameter error vector, and $\omega(t) = \left[\theta_4^h, \theta_4^h s(t), \theta_4^h s^2(t), \theta_4^h s^3(t), 1\right]^T$ is a regressor.

In summary, for both cases of an exact jet model and an approximate model, the proposed adaptive inverse schemes have led to linearly parameterized control error expressions which are crucial for the development of adaptive update laws. The design of such adaptive laws requires a feedback control law for generating the signal $C_{Md}(t)$. In more recent works, researchers are working on advanced control techniques for incorporating SJs to eliminate the conventional rudder section of the wind for performing roll control at high AoA. In a recently published work, Liu et al. have proposed a finite-time decoupling control Super-twisting sliding mode control strategy for aircraft with thrust vector at high AoA manoeuvre to account for the insufficient aerodynamic control surface to accomplish expected performance and stability [24]. In the next chapter, we explore important control challenges when the AoA changes over a larger space from (low AoA to high AoA) [25], thereby requiring a novel control methodology.

References

1. Erickson, G.: High angle-of-attack aerodynamics. Ann. Rev. Fluid Mech. **27**(1), 45–88 (1995). https://doi.org/10.1146/annurev.fl.27.010195.000401
2. Kelly, B.: Beyond the basics of ICING. Aviat. Saf. **1** 3(22), 1–6 (1993)
3. Linehan, T., Mohseni, K., Ferrar, J.: Active roll control at high angles of attack via bio-inspired sliding alula. AIAA Scitech 2020 Forum (2020). https://doi.org/10.2514/6.2020-1541
4. Post, M., Corke, T.: Separation control on high angle of attack airfoil using plasma actuators. AIAA J. **42**(11), 2177–2184 (2004). https://doi.org/10.2514/1.2929
5. Atkinson, M., Poggie, J., Camberos, J.: Control of high-angle-of-attack reentry flow with plasma actuators. J. Spacecr. Rocket. **50**(2), 337–346 (2013). https://doi.org/10.2514/1.a32360
6. Zheng, B., Xue, M., Ge, C.: Sliding discharge plasma actuation for forebody vortex control on a slender body at high angles of attack. AIP Adv. **10**(5) (2020). https://doi.org/10.1063/5.0005366
7. Zong, H., Chiatto, M., Kotsonis, M., de Luca, L.: Plasma synthetic jet actuators for active flow control. Actuators **7**(4), 77 (2018). https://doi.org/10.3390/act7040077
8. Fears, S.P., Ross, H.M., Moul, T.M.: Low-Speed wind-tunnel investigation of the stability and control characteristics of a series of flying wings with sweep angles of 50°. NASA Tech. Memo. **4640**
9. Kondor, S., Parekh, D., Washburn, A., Glezer, A.: Investigations of synthetic jet aerodynamic control modulation on a full-scale UAV. In: 43rd AIAA Aerospace Sciences Meeting And Exhibit (2005). https://doi.org/10.2514/6.2005-1051
10. Walter, D., Watkins, T.: Force and pitching moment variation of inverted aerofoils at high angles of attack in ground effect. In: 16th Australasian Fluid Mechanics Conference Crown Plaza, Gold Coast, Australia (2007)
11. Deb, D., Tao, G., Burkholder, J., Smith, D.: An adaptive inverse control scheme for synthetic jet actuator arrays. Infotech@Aerospace (2005). https://doi.org/10.2514/6.2005-7170
12. Farrell, J., Sharma, M., Polycarpou, M.: Backstepping-Based flight control with adaptive function approximation. J. Guid. Control Dyn. **28**(6), 1089–1102 (2005). https://doi.org/10.2514/1.13030
13. Liang, Y., Cong, S., Shang, W.: Function approximation-based sliding mode adaptive control. Nonlinear Dyn. **54**(3), 223–230 (2008). https://doi.org/10.1007/s11071-007-9324-0
14. Hofinger, A.: Nonlinear function approximation: computing smooth solutions with an adaptive greedy algorithm. J. Approx. Theory **143**(2), 159–175 (2006). https://doi.org/10.1016/j.jat.2006.03.016
15. Deb, D., Tao, G., Burkholder, J., Smith, D.: An adaptive inverse control scheme for a synthetic jet actuator model. In: Proceedings of the 2005, American control conference (2005). https://doi.org/10.1109/acc.2005.1470367
16. Deb, D., Tao, G., Burkholder, J., Smith, D.: Adaptive synthetic jet actuator compensation for a nonlinear tailless aircraft model at low angles of attack. In: 2006 American Control Conference (2006). https://doi.org/10.1109/acc.2006.1655376
17. Deb, D., Tao, G., Burkholder, J., Smith, D.: Adaptive synthetic jet actuator compensation for a nonlinear aircraft model at low angles of attack. IEEE Trans. Control Syst. Technol. **16**(5), 983–995 (2008). https://doi.org/10.1109/tcst.2007.912124
18. Devaud, E., Siguerdidjane, H., Font, S.: Some control strategies for a high-angle-of-attack missile autopilot. Control Eng. Pract. **8**(8), 885–892 (2000). https://doi.org/10.1016/s0967-0661(00)00013-7
19. Pedro, J., Mansfield, B.: Nonlinear control of aircraft at high angles of attack. In: 2004 IEEE Africon. 7th Africon Conference In Africa. https://doi.org/10.1109/africon.2004.1406711
20. de Freitas Virgilio Pereira, M., Prado, I., de Castro, D., Balthazar, J., da Silva, R., Nabarrete, A.: On nonlinear dynamics and flight control at high angles of attack with uncertain aerodynamics, vol. 4B. Dynamics, Vibration, and Control (2016). https://doi.org/10.1115/imece2016-67108

21. Deb, D., Tao, G., Burkholder, J.: Adaptive approximation based compensation of synthetic jet actuator nonlinearities for aircraft control. AIAA-2007-6523. In: Proceedings of the 2007 AIAA Guidance, Navigation and Control Conference, Hilton Head, SC (2007)
22. Hellman, M.: A graphical determination of the nature of the roots of a cubic. Math. Mag. **34**(4), 221–222 (1961). https://doi.org/10.1080/0025570x.1961.11975224
23. O'Connell, J., Haile, J.: Thermodynamics: P Fundamentals for Applications (2005). https://doi.org/10.1017/cbo9780511840234
24. Liu, J., Sun, M., Chen, Z., Sun, Q.: Super-twisting sliding mode control for aircraft at high angle of attack based on finite-time extended state observer. Nonlinear Dyn. **99**, 2785–2799 (2020). https://doi.org/10.1007/s11071-020-05481-1
25. Deb, D., Tao, G., Burkholder, J.: An adaptive inverse compensation scheme for signal-dependent actuator nonlinearities. In: 2007 46th IEEE Conference on Decision and Control (2007). https://doi.org/10.1109/cdc.2007.4434743

Chapter 5
Signal-Dependent Uncertainty Compensation: A General Framework

The process of modeling of SJAs at low AoA [1, 2] is very different from that at high AoA [3]. The structure of the actuator nonlinearity is also very different. Therefore, while deploying parametrized nonlinear functions for the SJAs, there must be a consideration for accommodating an appropriate function that is suitable for a reasonable range of AoA of an air vehicle. The characteristics of SJAs at low AoA and high AoA are significantly different as has been seen in the earlier chapters as well. As a consequence, a unified modeling framework for SJAs that is valid for the entire operating regime of AoA results in the actuator nonlinearity function being dependent on the AoA. That is, the SJA nonlinearity function varies with the AoA: a state of the aircraft, making the actuator function a state-dependent or signal-dependent [4]. Such SJA nonlinearities, in the topics dealt so far in the book, are significantly dependent only on the AoA among the states of the aircraft. We conceive that other states of the aircraft can also affect the SJA uncertainties in a nonlinear sense. However, wind tunnel tests conducted with SJAs at leading and trailing edge of airfoils have so far not revealed such a specific dependency. Therefore, a generalized class of systems with the actuators representing the dependencies over multiple signals which is a hitherto unaddressed problem in the literature needs to be systematically studied.

In this chapter, we present a different adaptive inversion mechanism utilizing two approximators. One of them approximates the actuator nonlinearity function, and the other handles the associated adaptive inverse formulation. At the outset, we formulate a generalized linear parametrization representing the approximator for an actuator model with multiple-signal dependencies. We also address the adaptive compensation formulation for such nonlinearities, followed by an adaptive inverse controller that facilitates a state feedback control signal. Subsequently, we present an adaptive framework to estimate the parameters of the two nonlinearity approximators which are coupled to each other. The approximators adapt to the uncertainties of each other and over time compensate for the same in such a way that the control error is

D. Deb et al., *Adaptive Compensation of Nonlinear Actuators for Flight Control Applications*, Studies in Systems, Decision and Control 386,
https://doi.org/10.1007/978-981-16-4161-9_5

diminished. We also innovatively accommodate the residual error term and develop appropriate adaptive laws to adjust the parameters. The proposed control design provides guaranteed tracking within desired boundaries.

5.1 Signal-Dependent Nonlinearity Compensation

Many actuators have state-dependant input-output characteristics [5]. There are also classes of nonlinear systems with state-dependent actuator parametric bounds and bias [6]. With a generic formulation of multiple-signal dependencies in actuator nonlinearities, we study the model

$$u = N(\phi^*; v, x_s), \tag{5.1}$$

where u is the actual output from the actuators which is not measurable, v is the actuator input vector, x_s represents the states contributing to the actuator uncertainties, and ϕ^* is a vector containing uncertain parameters. The objective is to find a parametrization of $N(\cdot)$ appropriate for adaptive compensation.

We consider a class of nonlinear systems given by

$$\begin{aligned} \dot{x}_i &= f_{x_i} + g_{x_i}\bar{u_i}, i = 1, \ldots, k, \\ \dot{x}_{k+1} &= f_{x_{k+1}} + g_{x_{k+1}}u + h_{x_{k+1}}, \end{aligned} \tag{5.2a}$$

$$\vdots$$

$$\begin{aligned} \dot{x}_{n-m} &= f_{x_{n-m}} + g_{x_{n-m}}h_{x_{n-m}}, \\ \dot{x}_{n-m+1} &= f_{x_{n-m+1}} + g_{x_{n-m+1}}u, \end{aligned} \tag{5.2b}$$

$$\vdots$$

$$\dot{x}_n = f_{x_n} + g_{x_n}u, \tag{5.2c}$$

$$u = N(v, x_s), \ u, v \in R^m, \tag{5.2d}$$

where m represents the number of actuators, $x_s \subset x = [x_1, x_2, \ldots, x_n]^T$, and f_{x_i} and g_{x_i} are the known functions in the ith equation, and h_{x_i} are the functions that are at least partially unknown. For continuous functions $f(v, x_s), v, x_s \in \mathrm{S} \subset R^n$, a compact set S, the nonlinearity model is best expressed by one of these approximators under different conditions and have been extensively studied in literature.

Approximator architectures such as Neural networks, Taylor series approximators etc. aim to find an arbitrarily complex mathematical function $y = f(\cdot)$ that can accurately map the attributes to the output, depending on the dataset and the architecture itself. The Universal Approximation Theorem purports that such approximators have a kind of *universality*, that is, irrespective of $f(\cdot)$, there is a network that can approximately approach the result for any number of inputs and outputs [7]. Gardiner describe two conditions which when satisfied, ensure that the universality of Taylor series holds [8]. We express the actuator nonlinearity as

$$u = N(v, x_s) \triangleq f(v, x_s) = \phi_1^{*T} \omega_1(v, x_s) + \epsilon_1(v, x_s), \tag{5.3}$$

where $x_s = [x_{s1}, x_{s2}, \ldots, x_{sM},]$, and ϕ_1^* represents unknown parameters given as

$$\phi_1^* = \Bigg[f(p, q), \frac{\partial(v, x_s)}{\partial v}|_{p,q}, \frac{\partial(v, x_s)}{\partial x_{s1}}|_{p,q}, \ldots, \frac{\partial(v, x_{s1})}{\partial x_{sM}}|_{p,q}, \ldots, \frac{\partial^{R_v}}{\partial v^{R_v}} \frac{\partial^{R_{x_s}}}{\partial x_{s1}^{R_{x_s}}}, \ldots,$$
$$\frac{\partial^{R_{x_{sM}}}}{\partial x_{sM}^{R_{x_{sM}}}} f(v, x_s)|_{p,q} \Bigg]^T \epsilon\ R^{(R_v+1)(R_{x_s}+1),\ldots,(R_{x_{sM}}+1)} \tag{5.4}$$

with $q = [q_1, q_2, \ldots, q_M]$, and $\epsilon_1(v, x_s)$ is the error term representing the modeling inaccuracies, and $\omega_1(v, x_s)$ is known vector function given by

$$\omega_1 = \Bigg[1, \frac{v-p}{1!}, \frac{x_{s1}-q_1}{1!}, \frac{x_{s2}-q_2}{1!}, \ldots, \frac{x_{sM}-q_M}{1!}, \ldots,$$
$$\frac{(v-p)^{R_v}(x_{s1}-q_1)^{R_{x_{s1}}}, \ldots, (x_{sM}-q_M)^{R_{x_{sM}}}}{R_v! R_{x_{s1}}! R_{x_{s2}}! R_{x_{sM}}!} \Bigg]^T, \tag{5.5}$$

with suitable integers R_v, $R_{x_{s1}}$, $R_{x_{s2}}$, ..., $R_{x_{sM}}$, and real scalars p, q_1, q_2, ..., q_M and r_1, r_2, ..., r_{M+1} such that $|v - p| < r_1$, $|x_{si} - q_i| < r_{i+1}, i = 1, \ldots, M$. The modeling error is reduced with choice of larger values of R_v, $R_{x_{s1}}$, $R_{x_{s2}}$,..., $R_{x_{sM}}$. The Taylor series approximation is

$$f(v, x_s) = \sum_{r_v=0}^{R_v} \sum_{r_{x_{s1}}=0}^{R_{x_{s1}}} \cdots \sum_{r_{x_{sM}}=0}^{R_{x_{sM}}} \frac{1}{r_v! r_{x_{s1}}! \ldots r_{x_{sM}}!} \frac{\partial^{R_v}}{\partial v^{R_v}} \frac{\partial^{R_{x_{s1}}}}{\partial x_{s1}^{R_{x_{s1}}}}$$
$$\cdots \frac{\partial^{R_{x_{sM}}}}{\partial x_{sM}^{R_{x_{sM}}}} f(v, x_s)|_{p,q} (v-p)^{R_v} (x_{s1}-q_1)^{R_{x_{s1}}} \ldots (x_{sM}-q_M)^{R_{x_{sM}}},$$

where $\frac{\partial^{R_{x_{s1}}}}{\partial x_{s1}^{R_{x_{s1}}}} \ldots \frac{\partial^{R_{x_{sM}}}}{\partial x_{sM}^{R_{x_{sM}}}} f(v, x_s)|_{p,q}$ are smooth nonlinear and invertible functions, but unknown so that compensation is difficult. For a parametrized nonlinearity function

$\bar{u} = \phi_1^{*T}\omega_1(v, x_s)$, an estimate $\hat{u}$ is formulated as

$$\hat{u} = \hat{N}(v, x_s) = \hat{\phi}_1^T\omega_1(v, x_s). \tag{5.6}$$

5.1.1 Adaptive Inverse Function

With an inverse characteristic such that $u = N(NI(u_d, x_s), x_s) = u_d$ for a desired signal u_d, we employ a Taylor series approximator for functional estimation of

$$u_{dT} = NI(u_d, x_s) \triangleq f^{-1}(u_d, x_s) - u_d, \tag{5.7}$$

where $f^{-1}(u_d, x_s)$ is the inverse function for the SJA nonlinearity. A modified non-linearity inverse function u_{dT} is expressed as

$$u_{dT} = NI(u_d, x_s) \triangleq g(u_d, x_s) = \phi_2^{*T}\omega_2(u_d, x_s) + \epsilon_2(u_d, x_s), \tag{5.8}$$

where ϕ_2^* is an unknown parameter vector, $\epsilon_2(u_d, \alpha)$ is the modeling error, and $\omega(u_d, \alpha)$ is a known regressor

$$\omega_2 = \left[1, \frac{u_d - c}{1!}, \frac{x_{s1} - q_1}{1!}, \frac{x_{s2} - q_2}{1!}, \ldots, \frac{x_{sM} - q_M}{1!}, \ldots, \right.$$
$$\left. \frac{(u_d - c)^{R_v}(x_{s1} - q_1)^{R_{xs1}}, \ldots, (x_{sM} - q_M)^{R_{xsM}}}{\bar{R}_v!\bar{R}_{x_{s1}}!\bar{R}_{x_{s2}}!\bar{R}_{x_{sM}}!}\right]^T \tag{5.9}$$

with appropriate natural numbers $\bar{R}_v$, $\bar{R}_{x_{s1}}$, $\bar{R}_{x_{s2}}$, ..., $\bar{R}_{x_{sM}}$, and real scalars c, q_1, q_2, ..., q_M and $\bar{r}_1, r_2, \ldots, r_{M+1}$ such that $|u_d - c| < \bar{r}_1, |x_{si} - q_i| < r_{i+1}, i = 1, \ldots, M$. The approximator is

$$g(u_d, x_s) = \sum_{\bar{r}_v=0}^{\bar{R}_v}\sum_{\bar{r}_{x_{s1}}=0}^{\bar{R}_{x_{s1}}}\cdots\sum_{\bar{r}_{x_{sM}}=0}^{\bar{R}_{x_{sM}}}\frac{1}{\bar{r}_v!\bar{r}_{x_{s1}}!\ldots\bar{r}_{x_{sM}}!}\frac{\partial^{\bar{R}_v}}{\partial v^{\bar{R}_v}}\frac{\partial^{\bar{R}_{x_{s1}}}}{\partial x_{s1}^{\bar{R}_{x_{s1}}}}\cdots$$
$$\frac{\partial^{\bar{R}_{x_{sM}}}}{\partial x_{sM}^{\bar{R}_{x_{sM}}}}f(v, x_s)|_{c,q}(u_d - c)^{\bar{R}_v}(x_{s1} - q_1)^{\bar{R}_{x_{s1}}}\ldots(x_{sM} - q_M)^{\bar{R}_{x_{sM}}},$$

where $\frac{\partial^{\bar{R}_{x_{s1}}}}{\partial x_{s1}^{\bar{R}_{x_{s1}}}}, \ldots, \frac{\partial^{\bar{R}_{x_{sM}}}}{\partial x_{sM}^{\bar{R}_{x_{sM}}}}f(v, x_s)|_{c,q}$ are smooth and invertible. For the parametrization $\bar{u}_{dT} = \phi_2^{*T}\omega_2(u_d, x_s)$, an estimate $\hat{u}_{dT}$ can be expressed as

$$\hat{u}_{dT} = \widehat{NI}(u_d, x_s) = \hat{\phi}_2^T\omega_2(u_d, x_s). \tag{5.10}$$

Using (5.7) and $\tilde{u}_{dT} = u_{dT} - \hat{u}_{dT}$, we get

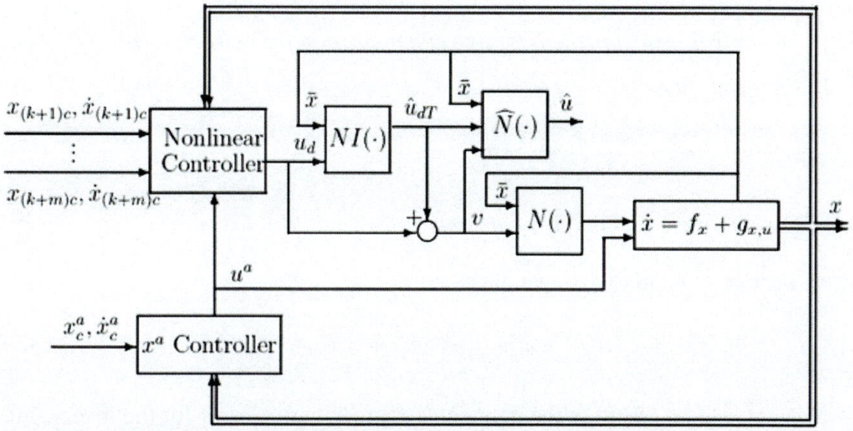

Fig. 5.1 Twin approximator-based adaptive compensation system

$$u_d = f(u_d + \hat{u}_{dT} + \tilde{u}_{dT}, x_s). \tag{5.11}$$

Figure 5.1 presents a twin approximator-based inverse compensation control framework using (5.3), (5.6), and (5.10) as the actuation function and the two approximation functions respectively. The external inputs are $\bar{u}_i, i = 1, \ldots, k$.

5.1.2 Control Error Expression

By expanding the Taylor series about $u_{dT} = \hat{u}_{dT}$, and using applied control

$$v = u_d + \hat{u}_{dT}, \tag{5.12}$$

we have u_d expressed as

$$u_d = f(v, x_s) + \frac{\partial f(v, x_s)}{\partial v}\tilde{u}_{dT} + O\left(\frac{\partial f^i(v, x_s)}{\partial v}, \tilde{u}_{dT}\right), \tag{5.13}$$

where $i = 2, 3, \ldots$, and $O(\cdot)$ represents the sum of the higher order arguments. The importance of this equation is that it replaces u_d which is nonlinear in $\tilde{u}_{dT}$ by an expression linear in $\tilde{u}_{dT}$ plus higher-order terms.

We develop a control error expression to demonstrate the effectiveness of our coupled nonlinearity function approximation. We make an assumption that

(A1) ϕ_1^* and ϕ_2^* are bounded, that is, $||\phi_1^*|| \leq \phi_A$ and $||\phi_2^*|| \leq \phi_C$ where ϕ_A and ϕ_C are known, and the modelling errors $\epsilon_1(v, x_s)$ and $\epsilon_2(u_d, x_s)$ are bounded such that $||\epsilon_1|| \leq \epsilon_{N2}$.

Lemma 5.1 *Under assumption (A1), with the compensator (5.10) and the observer (5.6), the control error for the adaptive compensation framework involving SJAs, is given by*

$$u - u_d = \tilde{\phi}_1^T \frac{\partial \omega_1(v, x_s)}{\partial v} \hat{\phi}_2^T \omega_2(u_d, x_s) - \hat{\phi}_1^T \frac{\partial \omega_1(v, x_s)}{\partial v} \tilde{\phi}_2^T \omega_2(u_d, x_s) + \epsilon, \quad (5.14)$$

where ϵ is a bounded model mismatch error.

Proof From (5.8) and (5.10), we have

$$\tilde{u}_{dT} = \tilde{\phi}_2^T \omega_2(u_d, x_s) + \epsilon_2(u_d, x_s), \quad (5.15)$$

where $\tilde{\phi}_2(t) = \phi_2^* - \hat{\phi}_2(t)$ is the parameter approximation error for the inverse non-linearity function. From (5.3) and (5.13), the control error is

$$u - u_d = -\frac{\partial f(v, x_s)}{\partial v} \tilde{u}_{dT} - O\left(\frac{\partial f^i(v, x_s)}{\partial v}, \tilde{u}_{dT}\right). \quad (5.16)$$

Substituting (5.3) and (5.13) in (5.15), we get

$$u - u_d = -\frac{\partial f(v, x_s)}{\partial v} \tilde{\phi}_2^T \omega_2(u_d, x_s) - \frac{\partial f(v, x_s)}{\partial v} \epsilon_2(u_d, x_s) - O\left(\frac{\partial f^i(v, x_s)}{\partial v^i}, \tilde{u}_{dT}\right).$$

For parameter approximation error $\tilde{\phi}_1(t) = \phi_1^* - \hat{\phi}_1(t)$, we get

$$u - u_d = \tilde{\phi}_1^T \frac{\partial \omega_1(v, x_s)}{\partial v} \hat{\phi}_2^T \omega_2(u_d, x_s) - \hat{\phi}_1^T \frac{\partial \omega_1(v, x_s)}{\partial v} \tilde{\phi}_2^T \omega_2(u_d, x_s) + \epsilon, \quad (5.17)$$

where ϵ is a bounded model mismatch error given by

$$\epsilon = -\tilde{\phi}_1^T \frac{\partial \omega_1(v, x_s)}{\partial v} \phi_2^{*T} \omega_2(u_d, x_s) - \frac{\partial \epsilon_1(v, x_s)}{\partial v} \tilde{\phi}_2^T \omega_2(u_d, x_s) - \frac{\partial \epsilon_1(v, x_s)}{\partial v} \epsilon_2(u_d, x_s)$$
$$-\phi_1^{*T} \frac{\partial \omega_1(v, x_s)}{\partial v} \epsilon_2(u_d, x_s) - O\left(\frac{\partial f^i(v, x_s)}{\partial v}, \tilde{u}_{dT}\right). \quad (5.18)$$

The control error expression (5.17) provides some interesting insights. The parameter error vectors and the parameter estimates are contained in both the non-modeling error terms.

5.1.3 Estimation of the Model Mismatch Term

The remainder term $O\left(\frac{\partial f^i(v,x_s)}{\partial v}, \tilde{u}_{dT}\right)$ quantifies deviation from the nominal function, and for a first order Taylor expansion, the inequality

$$\begin{aligned}\left\| O\left(\frac{\partial f^i(v,x_s)}{\partial v}, \tilde{u}_{dT}\right)\right\| &\leq \frac{1}{2}\left\|\frac{\partial^2 \epsilon_1(v,x_s)}{\partial v^2} + \phi_1^{*T}\frac{\partial^2 \omega_1(v,x_s)}{\partial v^2}\right\| \\ &\quad \left\|\tilde{\phi}_2^T \omega_2(u_d,x_s) + \epsilon_2(u_d,x_s)\right\|^2 \\ &\leq \delta_1||\hat{\phi}_2||^2 + \delta_5||\hat{\phi}_2|| + \delta_6, \end{aligned} \tag{5.19}$$

is satisfied where the second inequality is obtained by using the assumptions on the bounds of the nominal parameters ϕ_1^* and ϕ_2^*, and $||\tilde{\phi}_1|| \leq \phi_A + ||\hat{\phi}_1||$, $||\tilde{\phi}_2|| \leq \phi_C + ||\hat{\phi}_2||$, such that

$$\begin{aligned}|\epsilon(t)| &\leq \phi_C(\phi_A + ||\hat{\phi}_1||)\left\|\frac{\partial \omega_1(v,x_s)}{\partial v}\right\| ||\omega_2(u_d,x_s)|| + \phi_A \left\|\frac{\partial \omega_1(v,x_s)}{\partial v}\right\| \epsilon_{N2} \\ &+ \left|\frac{\partial \epsilon_1(v,x_s)}{\partial v}\right| \epsilon_{N2} + \left|\frac{\partial \epsilon_1(v,x_s)}{\partial v}\right| (\phi_C + ||\hat{\phi}_2||)||w_2(u_d,x_s)|| \\ &+ \delta_1||\hat{\phi}_2||^2 + \delta_5||\hat{\phi}_2|| + \delta_6, \end{aligned} \tag{5.20}$$

where δ_i, $i = 1, 5, 6$ are bounded but not necessarily known. Note that the bounds ϕ_1 and ϕ_2 of the parameter vectors ϕ_1^* and ϕ_2^* are not assumed to be known. With some algebraic simplifications, we get

$$|\epsilon(t)| \leq \delta_1 + \delta_2||\hat{\phi}_1|| + \delta_3||\hat{\phi}_2|| + \delta_4||\hat{\phi}_2||^2 = \xi^T \Phi, \tag{5.21}$$

where $\xi = [\delta_1, \delta_2, \delta_3, \delta_4]^T$ is a bounded constant vector estimated by projection based adaptive laws for $\hat{\xi} = [\hat{\delta}_1, \hat{\delta}_2, \hat{\delta}_3, \hat{\delta}_4]^T$, and $\Phi = [1, ||\hat{\phi}_1||, ||\hat{\phi}_2||, ||\hat{\phi}_1||^2]^T$ is known. Note that the control error (5.17) reflects the mutual dependence of the two parametrized nonlinearity functions.

We define an estimator for the bound of the model mismatch error $\hat{\xi}^T \Phi$ as

$$\hat{\epsilon} = \hat{\xi}^T \omega. \tag{5.22}$$

As the estimated terms $\hat{\phi}_1^T$ and $\hat{\phi}_2^T$ move asymptotically closer to the true parameters ϕ_1^{*T} and ϕ_2^{*T}, the Taylor series approximator effectively inverts the uncertainty function. The modification of one nonlinearity function depends on the other, and therefore the adaptive update laws to be developed for $\hat{\phi}_1^T$ and $\hat{\phi}_2^T$ should be mutually coupled, such that the parameter errors $\tilde{\phi}_1^T$ and $\tilde{\phi}_2^T$ are bounded and closed-loop stability is ensured. An explicit expression for ξ is not needed to be known and are estimated through an adaptive update law.

5.2 Adaptive Inverse Control System

The adaptive control design is inherently nonlinear and the analysis and design rely heavily on Lyapunov stability theory. Next, we describe an adaptive inverse compensation framework with online parameter adaptation for cancellation of $N(\phi^*; v, x_s)$. The control inputs (actuator outputs) are $u = N(v, x_s) = [u_1, u_2, \ldots, u_m]^T$, and external inputs $\bar{u}$ control states $x^a \subset x$ such that $x^a \cap \bar{x} = \phi$. The control law will also depend on intermediate states such that $x^a \cup x^b = x, x^a \cap x^b = \phi$ [9–12].

We consider that the states $x^q = (x_1, ..., x_k)$ are governed by external inputs, and the remaining subsets of equal number of states are given by $(x_{k+1}, \ldots, x_{k+m})$, $(x_{k+m+1}, \ldots, x_{k+2m})$,...,$(x_{n-m+1}, \ldots, x_n)$. This distribution is based on specific form of the g_{x_m}, u terms [13]. The desired trajectory is indicated by the bounded and differentiable reference signals $x_{ic}, i = 1, 2, \ldots, k+m$. This design is extendable to classes of nonlinear systems with g_{x_k} unknown.

External control development: From the x^a dynamics, the desired input $\bar{u}_i$ as a control command is

$$\bar{u}_i = \frac{1}{g_{x_i}}(-f_{x_i} + \dot{x}_{ic} - K_{x_i}\tilde{x}_i), i = 1, \ldots, k, \tag{5.23}$$

where x_{ic} represents reference signal, $\tilde{x}_i = x_i - x_{ic}$ denotes error states, $K_{x_i} > 0$, are constant gains.

Using (5.2a) and (5.23), the error dynamics is given by

$$\dot{\tilde{x}}_i = -K_{x_i}\tilde{x}_i, \ i = 1, \ldots, k. \tag{5.24}$$

Intermediate control formulations: For nonlinear dynamics

$$\dot{x}_r = f_{x_r} + g_{x_r}u + h_{x_{m+r}}, r = k+1, \ldots, k+m, \tag{5.25}$$

the control command generation is given by

$$h_{x_{(r+m)c}} + g_{x_r}(u_d + \hat{\xi}^T\Phi) = -f_{x_r} - K_{x_r}\tilde{x}_r + \dot{x}_{rc}, \tag{5.26}$$

where $K_{x_r} > 0, r = k+1, \ldots, k+m$, are control gains.

From (5.26), the functions $h_{x_{(r+m)c}}$ can be solved to generate the reference states $x_{(r+m)c}$ to control the next subset of states. Note that $\hat{\xi}^T\Phi$ term is added for ensuring stability in the Lyapunov sense. It is important to note that a solvability region exists for $x_{(r+m)c}$ which is not exclusively due to usage of a certain type of actuators, but is inherently present in the nonlinear class of dynamic systems for any set of actuators.

From (5.25) and (5.26), the state error equations are given by

$$\dot{\tilde{x}}_r = -K_{x_r}\tilde{x}_r + g_{x_r}(u - u_d - \hat{\xi}^T\Phi) + h_{x_{r+m}} - h_{x_{(r+m)c}}, \ r = k+1, \ldots, k+m.$$

For nonlinear dynamics

$$\dot{x}_r = f_{x_r} + g_{x_r} u + h_{x_{r+m}}, \ r = k + m + 1, \ldots, k + 2m, \tag{5.27}$$

the control is generated from

$$h_{x_{(r+m)c}} + g_{x_r}(u_d + \hat{\xi}^T \Phi) = -f_{x_r} - K_{x_r}\tilde{x}_r + \dot{x}_{rc} + \tilde{l}_{x_{r-m}}. \tag{5.28}$$

Note that the function $\tilde{l}_{x_{r-m}}$ consists of terms with state errors.

Using (5.27) and (5.28), the error dynamics are

$$\begin{aligned}\dot{\tilde{x}}_r = & -K_{x_r}\tilde{x}_r + g_{x_r}(u - u_d - \hat{\xi}^T \Phi) + h_{x_{r+m}} - h_{x_{(r+m)c}} \\ & + \tilde{l}_{x_{r-m}}, r = k + m + 1, \ldots, k + 2m.\end{aligned} \tag{5.29}$$

$$\vdots$$

For nonlinear dynamics

$$\dot{x}_r = f_{x_r} + g_{x_r} u + h_{x_{r+m}}, \ r = n - 2m + 1, \ldots, n - m, \tag{5.30}$$

and the control is generated from

$$h_{x_{(r+m)c}} + g_{x_r}(u_d + \hat{\xi}^T \Phi) = -f_{x_r} - K_{x_r}\tilde{x}_r + \dot{x}_{kc} + \tilde{l}_{x_{r-m}}, \tag{5.31}$$

the error equations are

$$\dot{\tilde{x}}_r = -K_{x_r}\tilde{x}_r + g_{x_r}(u - u_d - \hat{\xi}^T \Phi) + h_{x_{r+m}} - h_{x_{(r+m)c}} + \tilde{l}_{x_{r-m}}.$$

For nonlinear dynamics

$$\dot{x}_r = f_{x_r} + g_{x_r} u, r = n - m + 1, \ldots, n, \tag{5.32}$$

and the control is generated from

$$g_{x_r}(u_d + \hat{\xi}^T \Phi) = -f_{x_r} - K_{x_r}\tilde{x}_r + \dot{x}_{rc} + \tilde{l}_{x_{r-m}}, \tag{5.33}$$

such that with $g_{x_r} u_d = \sum_{i=1}^{m} g_{x_{r_i}} u_{di}$, we have

$$\dot{\tilde{x}}_r = -K_{x_r}\tilde{x}_r + g_{x_r}(u - u_d - \hat{\xi}^T \Phi) + \tilde{l}_{x_{r-m}}.$$

Having formulated the error equations useful in the adaptive update laws, we choose such ordinary different equations for the estimates $\hat{\phi}_1(t)$ and $\hat{\phi}_2(t)$ with the objective of ensuring closed-loop signal boundedness, and specifically asymptotic boundedness of the tracking errors, that is, $\lim_{t\to\infty} ||\tilde{x}|| \leq \epsilon$, for an arbitrary ϵ.

5.3 Adaptation Scheme

Next, we employ a simple adjustment of "projecting" the estimates to ensure continuance in a compact convex set thereby yielding a robust adaptive controller and guaranteed parameter boundedness [9]. We make an assumption that the bounds of the nominal parameters ϕ_1^*, ϕ_2^* and ξ are known for some known constants $\phi_{1a}^b, \ldots, \phi_{k_1a}^b, \phi_{1a}^d, \ldots, \phi_{k_1a}^d > 0$, $k_1 = (R_1 + 1)(M_1 + 1)$, $\phi_{1c}^b, \ldots, \phi_{k_2c}^b$, $\phi_{1c}^d, \ldots, \phi_{k_2c}^d > 0$, $k_2 = (R_2 + 1)(M_2 + 1)$, $\delta_1^b, \ldots, \delta_4^b$ and $\delta_1^d, \ldots, \delta_4^d$. Initial estimates $\hat{\phi}_1(0)$, $\hat{\phi}_2(0)$ and $\hat{\xi}(0)$ of ϕ_1^*, ϕ_2^* and ξ respectively are chosen such that the parameter estimates are within the same bounds of the true parameters at all times.

Parameter Estimation Signals: We choose the adaptive laws as

$$
\begin{aligned}
\dot{\hat{\phi}}_1 &= g_1(t) + p_1(t),\\
\dot{\hat{\phi}}_2 &= g_2(t) + p_2(t),\\
\dot{\hat{\xi}}(t) &= \Gamma_3 \Phi |e(t)| + p_\xi(t),
\end{aligned}
\tag{5.34}
$$

where $g_1(t)$ and $g_2(t)$ are adaption functions given by

$$
\begin{aligned}
g_1(t) &= \Gamma_1 e(t) \frac{\partial \omega_1(v,\alpha)}{\partial v} \hat{\phi}_2^T \omega_2(u_d,\alpha),\\
g_2(t) &= \Gamma_2 e(t) \frac{\partial \omega_1(v,\alpha)}{\partial v} \hat{\phi}_1^T \omega_2(u_d,\alpha),
\end{aligned}
\tag{5.35}
$$

gain matrices $\Gamma_i = \Gamma_i^T > 0$, $i = 1, 2, 3$, and

$$
e(t) = \sum_{r=K+1}^{n} \tilde{x}_r g_{x_r},
\tag{5.36}
$$

and $p_1(t)$, $p_2(t)$ and $p_\xi(t)$ are projections so that the parameter estimates are constrained in a desired region, and are expressed as

$$
p_{ij}(t) = \begin{cases} 0 & \text{if } \hat{\phi}_{ij}(t) > \phi_{ij}^b, \text{ or} \\ & \text{if } \hat{\phi}_{ij}(t) = \phi_{ij}^b \text{ and } g_{ij}(t) \geq 0, \\ -g_{ij}(t) & \text{Otherwise,} \end{cases}
\tag{5.37}
$$

where $i = 1, \ldots, (R_1 + 1)(M_1 + 1)$, $j = a$, and $i = 1, \ldots, (R_2 + 1)(M_2 + 1)$, $j = c$, and

$$p_{\xi k}(t) = \begin{cases} 0 & \text{if } \hat{\delta}_k(t) \ \epsilon(\delta_k^b, \delta_k^d), \ \text{or} \\ & \text{if } \hat{\delta}_k(t) = \delta_k^b \ \ \Gamma_3 \Phi e(t) \le 0, \ \text{or} \\ & \text{if } \hat{\delta}_k(t) = \delta_k^d \ \ \Gamma_3 \Phi e(t) \ge 0, \\ \Gamma_3 \Phi e(t) & \text{Otherwise,} \end{cases} \tag{5.38}$$

where $k = 1, \ldots, 4$, $p_3 = [p_{\xi_1}, \ldots, p_{\xi_4}]^T$. Next, we present the stability analysis.

Theorem 5.1 *Under the assumptions specified for the parameter bounds, the error dynamics (5.24), (5.27), (5.29), (5.32) and (5.34) from state feedback control law (5.33), with the adaptive update laws (5.34) guarantee that the closed-loop signals are bounded and tracking errors are asymptotically bounded:*

$$\lim_{t \to \infty} \sup_{\tau \ge t} |\tilde{x}(t)| \le \epsilon,$$

where $\tilde{x} = [\tilde{x}_1, \ldots, \tilde{x}_n]^T$ [14–22].

Proof Let us consider a scalar continuous and positive definite function

$$V_L = \frac{1}{2}\tilde{x}^T \tilde{x} + \frac{1}{2}\tilde{\phi}_1^T \Gamma_1^{-1} \tilde{\phi}_1 + \frac{1}{2}\tilde{\phi}_2^T \Gamma_2^{-1} \tilde{\phi}_2 + \frac{1}{2}\tilde{\xi}^T \Gamma_3^{-1} \tilde{\xi}, \tag{5.39}$$

where $\tilde{\xi}(t) = \xi - \hat{\xi}(t)$. Using (5.24), (5.27), (5.29) and (5.34), and differentiating V_L and by cancelling some terms, we get

$$\begin{aligned} \dot{V}_L = & -\sum_{r=1}^{n} K_{x_r} \tilde{x}_r^2 + \sum_{r=K+1}^{n} \tilde{x}_r g_{x_r}(u - u_d - \hat{\xi}^T \Phi) + \sum_{r=K+1}^{n-m} \tilde{l}_{x_r} \\ & + \sum_{r=k+m+1}^{n} (h_{x_r} - \hat{h}_{x_r}) - \tilde{\phi}_1^T \Gamma_1^{-1} \dot{\hat{\phi}}_1 - \tilde{\phi}_2^T \Gamma_2^{-1} \dot{\hat{\phi}}_2 - \tilde{\xi}^T \Gamma_3^{-1} \dot{\hat{\xi}}. \end{aligned} \tag{5.40}$$

With an appropriate selection of $\tilde{l}_{x_r}$, we obtain

$$\sum_{r=k+m+1}^{n} (h_{x_r} - \hat{h}_{x_r}) + \sum_{r=k+1}^{n-m} \tilde{l}_{x_r} = \Delta(\tilde{x}_p) \le \kappa, \ \forall \tilde{x}_p \in D, \tag{5.41}$$

where $D = \{\tilde{x} | \ ||\tilde{x}|| \le \eta > 0\}$ is a compact set, and where x_p is a subset of x and $\Delta(\tilde{x}_p)$ are the higher order terms for small angles $\tilde{x}_p$. Substituting the control error (5.17) and adaptive laws (5.34) in (5.40), we have

$$\dot{V}_L = -\sum_{r=1}^{n} K_{x_r}\tilde{x}_r^2 + e(t)(u - u_d - \hat{\xi}^T\omega) + \Delta(\tilde{x}_p) - \tilde{\phi}_1^T\Gamma_1^{-1}\dot{\hat{\phi}}_1 - \tilde{\phi}_2^T\Gamma_2^{-1}\dot{\hat{\phi}}_2 - \tilde{\xi}^T\Gamma_3^{-1}\dot{\hat{\xi}}$$
$$-e(t)\hat{\xi}^T\Phi + e(t)\left(\tilde{\phi}_1^T\frac{\partial\omega_1(v,\alpha)}{\partial v}\hat{\phi}_2^T\omega_2(u_d,\alpha)\right) + e(t)\epsilon(t) - \xi^T\Phi|e(t)|$$
$$-e(t)\left(\hat{\phi}_1^T\frac{\partial\omega_1(v,\alpha)}{\partial v}\tilde{\phi}_2^T\omega_2(u_d,\alpha)\right) - \tilde{\xi}^T\Gamma_3^{-1}p_\xi(t)$$
$$= -\sum_{r=1}^{n} K_{x_r}\tilde{x}_r^2 + e(t)\epsilon(t) + \Delta(\tilde{x}_p) - \tilde{\phi}_1^T\Gamma_1^{-1}p_1(t) - \tilde{\phi}_2^T\Gamma_2^{-1}p_2(t) - \hat{\xi}^T\Phi e(t)$$
$$-\tilde{\xi}^T\Phi|e(t)| - \tilde{\xi}^T\Gamma_3^{-1}p_\xi(t). \tag{5.42}$$

With initial conditions the parametric bound condition and from the parameter projections $h_1(t)$ and $h_2(t)$ which maintain the parameters in the desired bounds, we find that the parametric bound condition is satisfied, and

$$\tilde{\phi}_1^T(t)h_1(t) \geq 0, \quad \tilde{\phi}_2^T(t)h_2(t) \geq 0, \quad \tilde{\xi}^T h_\xi(t) \geq 0. \tag{5.43}$$

Additionally, and since $|\epsilon(t)| \leq \xi^T\Phi$, (5.42) becomes

$$\dot{V}_L \leq -\sum_{r=1}^{n} K_{x_r}\tilde{x}_r^2 + |e(t)||\epsilon(t)| - \xi^T\Phi|e(t)| + \Delta(\tilde{x}_p) \leq -K||\tilde{x}(t)||_2^2 + \Delta(\tilde{x}_p), \tag{5.44}$$

where $K = \min(K_{x_r}, r = 1, \ldots, n)$. From (5.41), (5.44), signal boundedness requires

$$\frac{\kappa}{K} \leq \eta^2. \tag{5.45}$$

Signal Boundedness: Note that (5.45) is not the sufficient condition. First, we set the condition:

$$\sum_{i=1}^{3}\frac{\rho_i^2}{\lambda_{min}(\Gamma_i)} \leq \eta^2 - \frac{\kappa}{K}, \tag{5.46}$$

where we have assumed the conditions

$$||\tilde{\phi}_1(0)||_2 \leq \rho_1, \quad ||\tilde{\phi}_2(0)||_2 \leq \rho_2, \quad ||\tilde{\xi}(0)||_2 \leq \rho_3, \tag{5.47}$$

and when $||\tilde{x}(0)||_2 < \sqrt{\frac{\kappa}{K}}$, we get either $||\tilde{x}||_2 < \sqrt{\frac{\kappa}{K}}, t \geq 0$, or for $t \in [t_1, t_2]$, we have $\dot{V}_L \leq 0$, and $||\tilde{x}||_2 \leq \sqrt{\frac{\kappa}{K}}$ and we have

$$V_L(t) \leq V_L(t_1) \leq \sqrt{\frac{\kappa}{K}} + \sum_{i=1}^{3}\frac{\rho_i^2}{\lambda_{min}(\Gamma_i)} \leq \eta^2. \tag{5.48}$$

Therefore, $||\tilde{x}||_2 < \eta, \forall t \in [t_1, t_2]$. With a condition $||\tilde{x}(0)||_2 > \sqrt{\frac{\kappa}{K}}$, $\exists t_2 > 0$ so that $\dot{V}_L \leq 0, t \in [0, t_2]$. Then along with (5.46), we get

$$V_L(t) \leq V_L(t_1) \leq ||\tilde{x}(0)||_2^2 + \sum_{i=1}^{3} \frac{\rho_i^2}{\lambda_{min}(\Gamma_i)} \leq \eta^2. \tag{5.49}$$

Therefore, either $t_2 = \infty$ or there exists $t = t_3 > t_2$ such that $\dot{V}_L > 0, \forall t \in (t_2, t_3)$, that is, $||\tilde{x}||_2 < \sqrt{\frac{\kappa}{K}}$. Also, $||\tilde{x}||_2 \leq \eta, \forall t$ when the following condition is satisfied:

$$\max\left\{\sqrt{\frac{\kappa}{K}}, ||\tilde{x}(0)||_2^2\right\} + \sum_{i=1}^{3} \frac{\rho_i^2}{\lambda_{min}(\Gamma_i)} \leq \eta^2. \tag{5.50}$$

Therefore, all of the closed-loop signals are bounded, and we find convergence of the states to the compact set D ensuring semi-global stability. ∇

We have presented a twin approximator-based control structure operating with a coverage region, to cause the asymptotic tracking error to remain in a small neighborhood. Next, we evaluate the formulated theory using simulation results for an aircraft control application.

5.4 Performance Evaluation

For a simplified validation of the proposed approximation based compensation framework, we choose a high-fidelity and readily available aircraft model. For the study of controller effectiveness in a wide range of AoA, we define a tracking error $e(t)$ as the difference in the vertical acceleration and the desired value of the same state for an "ideal" aircraft with a linear dependency on the lift coefficient. The flight simulation condition is as follows: from a trim flight condition of approximately 135 knots and $\alpha = 2.5°$, a manual 50% column pulse of 15 s duration is inserted 1 s into the simulation.

The performance of the proposed controller using the twin Taylor series approximator framework, certainly, relies upon the chosen Taylor series attributes like the adaptation gains and initial parametric conditions, and the specific ones chosen are provided in Table 5.1.

An actuator model with nominal parameters $\phi^{h*} = \left[0, \ 0.02, \ \frac{\pi}{32}\right]$ is implemented as

$$\Delta C_L = \phi_1^{h*} + \phi_2^{h*} (\alpha - b)^2 \left(\phi_3^{h*} v\right). \tag{5.51}$$

In Fig. 5.2, three different cases are studied. In one case, represented in blue, the aircraft model operates without deploying the function for the SJAs and the corresponding inverse function, that is, with only the mechanical deflecting surfaces. In the second case, the SJ approximator applies on a fixed (non-adaptive) inverse,

Table 5.1 Twin Approximator-based Controller Parameters

Parameter	Value
q	16°
p, c	0
Γ_1, Γ_2	$0.1 \times I$ (I: Identity Matrix)
M_1, R_1	1
$\phi_2(0)$	$[0.05\ 0.05\ 0.05\ 0.05]^T$
$\phi_1(0)$	$[1\ 1\ 1\ 1]^T$

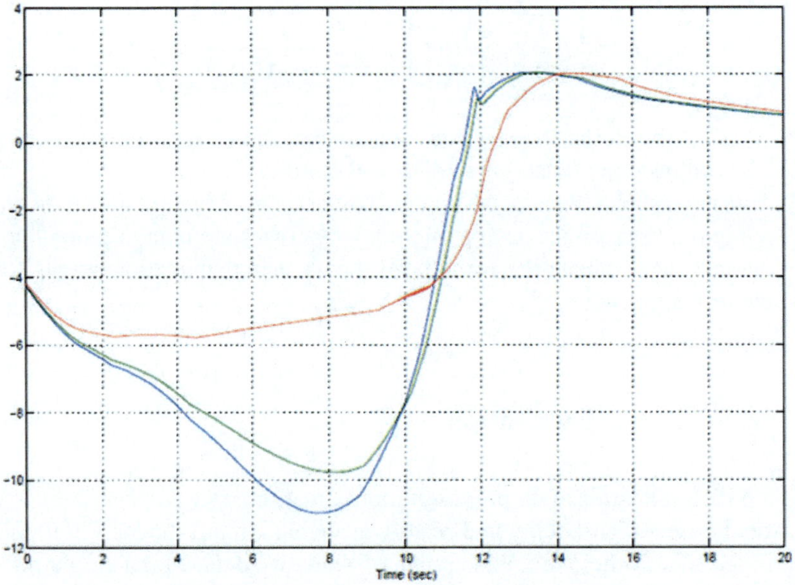

Fig. 5.2 Tracking errors: without SJs and without inverse (blue); with SJs and fixed inverse (green); and with jets and adaptive inverse (red)

as shown in green. Finally, in the third case, the twin approximator framework used for both the SJAs and the inverse function is shown in red. The tracking error in the third case is expectantly lowest among the three, and the first case has the maximum error. We model the SJs in the vicinity of the leading edge (apart from mechanical deflecting surfaces) to ensure stall is delayed and tracking error reduced. High AoA flight is always a safety and control issue, but the improved performance as per the simulation results show that it is feasible with the proposed arrangement. The transient performance relies on initial conditions, and control and adaptation gains.

References

1. Deb, D., Tao, G., Burkholder, J., Smith, D.: Adaptive synthetic jet actuator compensation for a nonlinear aircraft model at low angles of attack. IEEE Trans. Control Syst. Technol. **16**(5), 983–995 (2008). https://doi.org/10.1109/tcst.2007.912124
2. Deb, D., Tao, G., Burkholder, J., Smith, D.: An adaptive inverse control scheme for a synthetic jet actuator model. In: Proceedings of the 2005, American Control Conference (2005). https://doi.org/10.1109/acc.2005.1470367
3. Deb, D., Tao, G., Burkholder, J.: Adaptive approximation based compensation of synthetic jet actuator nonlinearities for aircraft control. AIAA-2007-6523. In: Proceedings of the 2007 AIAA Guidance, Navigation and Control Conference, Hilton Head, SC (2007)
4. Deb, D., Tao, G., Burkholder, J.: High-order design of adaptive inverses for signal-dependent actuator nonlinearities. In: 2008 American Control Conference (2008). https://doi.org/10.1109/acc.2008.4587029
5. Yamazaki, M., Yasunobu, S.: An intelligent control for state-dependent nonlinear actuator and its application to pneumatic servo system. In: SICE Annual Conference (2007). https://doi.org/10.1109/sice.2007.4421352
6. Zhao, Q.: Fault tolerant control against state-dependent actuator bias in a class of nonlinear systems. In: 2011 Chinese Control And Decision Conference (CCDC) (2011). https://doi.org/10.1109/ccdc.2011.5968624
7. Hornik, K.: Approximation capabilities of multilayer feedforward networks. Neural Netw. **4**(2), 251–257 (1991). https://doi.org/10.1016/0893-6080(91)90009-t
8. Gardiner, S.: Taylor Series, Universality and Potential Theory. Fields Institute Communications pp. 247–264 (2018). https://doi.org/10.1007/978-1-4939-7543-3_14
9. Naik, S., Kumar, P., Ydstie, B.: Robust continuous-time adaptive control by parameter projection. IEEE Trans. Autom. Control **37**(2), 182–197 (1992). https://doi.org/10.1109/9.121620
10. Bryson, A.E., Jr.: Control of Spacecraft and Aircraft. Princeton University Press, Princeton (1994)
11. Patel, R., Deb, D., Dey, R., Balas, V.E.: Introduction to adaptive control. Intelligent Systems Reference Library, pp. 53–65 (2019). https://doi.org/10.1007/978-3-030-18068-3_5
12. Deb, D., Tao, G., Burkholder, J., Smith, D.: Adaptive compensation control of synthetic jet actuator arrays for airfoil virtual shaping. J. Aircraft **44**(2), 616–626 (2007). https://doi.org/10.2514/1.24910
13. Deb, D., Tao, G., Burkholder, J.: An adaptive inverse compensation scheme for signal-dependent actuator nonlinearities. In: 2007 46th IEEE Conference on Decision and Control (2007). https://doi.org/10.1109/cdc.2007.4434743
14. Liu, Y., Deb, D., Tao, G.: Modeling and multivariable adaptive control of aircraft with synthetic jet actuators. In: 2008 7th World Congress on Intelligent Control and Automation (2008). https://doi.org/10.1109/wcica.2008.4593263
15. Deb, D., Sonowal, S.: Synthetic jet actuator based adaptive neural network control of nonlinear fixed pitch wind turbine blades. 2013 IEEE International Conference on Control Applications (CCA) (2013). https://doi.org/10.1109/cca.2013.6662759
16. Patel, R., Deb, D., Dey, R., Balas, V.E.: Model reference adaptive control of microbial fuel cells. Intelligent Systems Reference Library, pp. 109–121 (2019). https://doi.org/10.1007/978-3-030-18068-3_10
17. Patel, R., Deb, D., Dey, R., Balas, V.E.: Adaptive control of single chamber two-population MFC. Intelligent Systems Reference Library, pp. 81–89 (2019). https://doi.org/10.1007/978-3-030-18068-3_7
18. Patel, R., Deb, D., Dey, R., Balas, V.E.: Adaptive and intelligent control of microbial fuel cells. Intelligent Systems Reference Library (2020). https://doi.org/10.1007/978-3-030-18068-3
19. Nath, A., Deb, D., Dey, R.: An augmented subcutaneous type 1 diabetic patient modelling and design of adaptive glucose control. J. Proc. Control **86**, 94–105 (2020). https://doi.org/10.1016/j.jprocont.2019.08.010

20. Nath, A., Deb, D., Dey, R.: Robust observer based adaptive control of blood glucose in diabetic patients. Int. J. Control 1–14 (2020). https://doi.org/10.1080/00207179.2020.1750705
21. Deb, D., Tao, G., Burkholder, J., Smith, D.: Adaptive synthetic jet actuator compensation for a nonlinear tailless aircraft model at low angles of attack. In: 2006 American Control Conference (2006). https://doi.org/10.1109/acc.2006.1655376
22. Deb, D., Tao, G., Burkholder, J., Smith, D.: An adaptive inverse control scheme for synthetic jet actuator arrays. Infotech@Aerospace (2005). https://doi.org/10.2514/6.2005-7170

Chapter 6
NN-Based High-Order Adaptive Compensation Framework for Signal Dependencies

The adaptive inverse control technique effectively compensates for uncertain parameters in linear and nonlinear systems. The acquiescent characteristics of Neural networks based AIC for uncertain systems are ensuring much research interest in recent times. Researchers have attempted to solve the adaptive stabilization problem for a class of high-order nonlinear systems with inverse dynamics and nonlinear parameterization with partial state-feedback [1].

The main contributing aspects of this chapter are:

- An improved high-order inverse parametrization framework using NNs to improve approximation accuracy with signal-dependent actuator nonlinearities,
- An analytical study of this development to SJ actuation in flight control systems.

6.1 Higher-Order Inverse Parametrization Schemes

In this chapter, we perform adaptive compensation utilizing two NNs for higher-order approximations. At the outset, we should understand the underlying principle behind such an approximation process. We deploy the *Universal approximation theorem* for this purpose. The Universal Approximation Theorem is, very literally, the theoretical foundation of why NNs work. That is, it states that a NN with a single hidden layer and a finite but sufficient number of neurons can with reasonable accuracy approximate any continuous function, under certain conditions on the structure of the activation functions, namely, that they must be sigmoid-like.

Formulated in 1989 by George Cybenko specifically for sigmoidal activations [2] and proven by Kurt Hornik [3] in 1991 to apply to all activation functions, the discovery of universal approximation theory has been a significant driver in spurring the deployment of neural networks into a plethora of applications today.

D. Deb et al., *Adaptive Compensation of Nonlinear Actuators for Flight Control Applications*, Studies in Systems, Decision and Control 386,
https://doi.org/10.1007/978-981-16-4161-9_6

Most importantly, however, the theorem is an eye-opening explanation of why NNs appear to behave so intelligently. Understanding it is critical in developing a strong and deep understanding of NNs and the utilization in approximation based control.

One could construct a network by delegating each neuron to one approximation step using the weights and biases as zones in which an input falls and hence which neuron should be activated. A NN with a sufficient number of neurons could estimate a function simply be segregating it into several constant regions. For inputs that fall in a neuron's delegated section, by blowing the weight up to large values, the final value approaches 1 (when evaluated using the sigmoid function). If it does not fall into the section, moving the weight towards negative infinity will yield a final result near 0. The sigmoidal function approximates any given function with an abundance of neurons. Instead of creating complex mathematical relationships between the input and output, the universal approximation theorem uses simple linear manipulations to segregate the complicated function into many small, less complicated pieces, each of which are taken by one neuron. We should take care while training such networks to avoid over-fitting of the training data and then failure in generalization during new conditions.

Next, we present a new and improved adaptive parametrization scheme in presence of signal dependent actuator nonlinearities. We demonstrate this improved parametrization process in AIC through the higher-order components of the Taylor series expansion for NNs. This technique is however not limited to NN and any universal approximating tools such as fuzzy logic approximators, splines are also applicable. The effect of the signal dependence α is not a trivial extension of an actuator nonlinearity characteristic of the form $u = N(v)$. This is because, apart from an obvious extension of dependence in terms of a new dimension, the change of AoA in the actuation can even cause the system to be unstable.

A continuous function $f(v, \alpha), \forall v, \alpha \in S \subset R^2$, can be expressed in terms of parametric and non-parametric uncertainties for functional approximation, such that

$$u = N(v, \alpha) = f(v, \alpha) = W_1^T \phi(v, \alpha) + \epsilon_1(v, \alpha), \tag{6.1}$$

where $\phi(v, \alpha)$ is a radial basis function, W_1 is the constant outer-layer weight, and $\epsilon(v, \alpha)$ is the non-parametric (error) term [4].

Parametric uncertainties need a set of basis functions that can approximate a large class of functions within a given tolerance. For instance, Artificial Neural Networks (one such type of basis functions) are multi-input-multioutput systems composed of many interconnected nonlinear processing elements (neurons) operating in parallel.

For inner-layer weight V and outer-layer weight W_1, we use a basis $\phi(v, \alpha) = \sigma(V^T x_1 + v_0)$, and the actuator output is

$$u = W_1^T \sigma(V^T x_1 + v_0) + \epsilon_1(v, \alpha), \tag{6.2}$$

where $x_1 = [v, \alpha]^T$, and $\sigma(\cdot)$ is the hidden layer activation regressor, and v_0 are the thresholds (for a complete theory of NNs and their aircraft applications, see [5, 6]).

Function approximation using sigmoidal NN means finding connection weights W_1 and V to minimize $\epsilon_1(v, \alpha)$. These values are not necessarily unique. Typical selections for $\sigma(\cdot)$ include the sigmoid, polynomial basis functions and radial basis functions. For an estimate $\hat{W}_1$ for the nominal weight W_1, we have

$$\hat{u} = \hat{W}_1^T \sigma(V^T x_1 + v_0). \tag{6.3}$$

Only the second NN directly affects the control input. Ideally, for an inverse $v = NI(u_d, \alpha)$ and desired input signal u_d, we have

$$u = N(NI(u_d, \alpha), \alpha) = u_d. \tag{6.4}$$

We use another NN ($W_2^T \sigma_i(v_i^T x_2 + v_{0i})$) as a compensator, for estimation of u_{dNN}, expressed as

$$u_{dNN} = NI(u_d, \alpha) - u_d = W_2^T \sigma_i(V_i^T x_2 + v_{0i}) + \epsilon_2(u_d, \alpha), \tag{6.5}$$

where $\sigma_i(\cdot)$ represents hidden layer activation function, V_i and v_{0i} are basic NN elements, $x_2 = [u_d, \alpha]^T$ and $\epsilon_2(u_d, \alpha)$ is the non-parametric error, and the estimation is given as

$$\hat{u}_{dNN} = \hat{W}_2^T \sigma_i(V_i^T x_2 + v_{0i}), \tag{6.6}$$

where $\hat{W}_2$ is the estimation of the outer-layer weight W_2. The inner-layer weights V, V_i are randomly set and are fixed. From (6.2) and (6.4), we get

$$u_d = W_1^T \sigma\left(V^T [u_{dNN} + u_d, \alpha] + v_0\right) + \epsilon_1(v, \alpha), \tag{6.7}$$

Using (6.5) and (6.6), we get

$$\begin{aligned} u_d = W_1^T \sigma\Big(& V^T \left[\tilde{W}_2^T \sigma_i\left(V_i^T x_2 + v_{0i}\right) + \epsilon_2(u_d, \alpha), 0\right]^T \\ &+ V^T \left(x_2 + \left[\hat{u}_{dNN}, 0\right]^T\right) + v_0\Big) + \epsilon_1(v, \alpha), \end{aligned} \tag{6.8}$$

where $\tilde{W}_2 = W_2 - \hat{W}_2$ is a weight estimation error.

6.1.1 Neural Network Based Inverse Parametrization

For ideal outer-layer weights W_1 and W_2, we develop the error between the input and the desired input by making the following assumption:

(A1) for known bounds W_{1M} and W_{2M}, $||W_1||_F < W_{1M}$ and $||W_2||_F < W_{2M}$ W_{1M}.

Using Taylor series expansion at $(W_2, 0) = (\hat{W}_2, 0)$ in (6.8) up to second order, for estimation errors $\tilde{W}_1 = W_1 - \hat{W}_1$, $\tilde{W}_2 = W_2 - \hat{W}_2$, we get

$$\begin{aligned} u_d = {} & W_1^T \sigma\left(V^T\left(x_2 + [\hat{u}_{dNN}, 0]^T\right) + v_0\right) + \epsilon_1(v, \alpha) \\ & + W_1^T \sigma'\left(V^T\left(x_2 + [\hat{u}_{dNN}, 0]^T\right) + v_0\right) V^T \left[\tilde{W}_2^T \sigma_i\left(V_i^T x_2 + v_{0i}\right) + \epsilon_2(u_d, \alpha), 0\right]^T \\ & + W_1^T \sigma''\left(V^T\left(x_2 + [\hat{u}_{dNN}, 0]^T\right) + v_0\right)\left[\tilde{W}_2^T \sigma_i\left(V_i^T x_2 + v_{0i}\right) + \epsilon_2(u_d, \alpha), 0\right] \\ & .VV^T\left[\tilde{W}_2^T \sigma_i\left(V_i^T x_2 + v_{0i}\right) + \epsilon_2(u_d, \alpha), 0\right]^T + W_1^T R_2\left(\tilde{W}_2, u_d, \alpha\right), \end{aligned} \tag{6.9}$$

where $W_2^T R_2\left(\tilde{W}_2, u_d, \alpha\right)$ is the remainder of the second Taylor polynomial and σ' and σ'' are the first and second derivatives of $\sigma(x)$ with respect to x respectively. Since $x_1 = x_2 + [\hat{u}_{dNN}, 0]^T$, using (6.2), we get

$$u = W_1^T \sigma\left(V^T\left(x_2 + [\hat{u}_{dNN}, 0]^T\right) + v_0\right) + \epsilon_1(v, \alpha), \tag{6.10}$$

and by substituting (6.10) into (6.9), we get

$$\begin{aligned} u - u_d = {} & -W_1^T \sigma'\left(V^T x_1 + v_0\right) V^T\left[\tilde{W}_2^T \sigma_i\left(V_i^T x_2 + v_{0i}\right) + \epsilon_2(u_d, \alpha), 0\right]^T \\ & -\frac{1}{2} W_1^T \sigma''\left(V^T x_1 + v_0\right)\left[\tilde{W}_2^T \sigma_i\left(V_i^T x_2 + v_{0i}\right) + \epsilon_2(u_d, \alpha), 0\right] VV^T \\ & \cdot\left[\tilde{W}_2^T \sigma_i\left(V_i^T x_2 + v_{0i}\right) + \epsilon_2(u_d, \alpha), 0\right]^T - W_1^T R_2\left(\tilde{W}_2, u_d, \alpha\right). \end{aligned} \tag{6.11}$$

Next, utilizing the weighting errors $\tilde{W}_1 = W_1 - \hat{W}_1$, $\tilde{W}_2 = W_2 - \hat{W}_2$, we express (6.11) in a form that is linear in $\tilde{W}_1$ and $\tilde{W}_2$, and contains ϵ (nonlinear terms) and $\epsilon_2(u_d, \alpha)$ components. The control error signal $u - u_d$ is

$$\begin{aligned} u - u_d = {} & \tilde{W}_1^T \sigma'\left(V^T x_1 + v_0\right) V^T [\hat{u}_{dNN}, 0]^T - \hat{W}_1^T \sigma'\left(V^T x_1 + v_0\right) V^T \\ & \left[\tilde{W}_2^T \sigma_i\left(V_i^T x_2 + v_{0i}\right), 0\right]^T - \frac{1}{2} \tilde{W}_1^T \sigma''\left(V^T x_1 + v_0\right)[\hat{u}_{dNN}, 0] VV^T [\hat{u}_{dNN}, 0]^T \\ & + \frac{1}{2} \hat{W}_1^T \sigma''\left(V^T x_1 + v_0\right) V^T [\hat{u}_{dNN}, 0]^T \left[\tilde{W}_2^T \sigma_i\left(V_i^T x_2 + v_{0i}\right), 0\right] V + \epsilon \\ = {} & -\hat{W}_1^T \Psi\left[\tilde{W}_2^T \sigma_i\left(V_i^T x_2 + v_{0i}\right), 0\right]^T + \tilde{W}_1^T \Psi [\hat{u}_{dNN}, 0]^T + \epsilon, \end{aligned} \tag{6.12}$$

where Ψ is

$$\Psi = \left(\sigma'\left(V^T x_1 + v_0\right) - \frac{1}{2} \sigma''\left(V^T x_1 + v_0\right)[\hat{u}_{dNN}, 0] V\right) V^T, \tag{6.13}$$

and ϵ is

$$\epsilon = -W_1^T \sigma'\left(V^T x_1 + v_0\right) V^T \left[\epsilon_2(u_d, \alpha), 0\right]^T - \tilde{W}_1^T \sigma'\left(V^T x_1 + v_0\right) V^T$$
$$\left[\tilde{W}_2^T \sigma_i\left(V_i^T x_2 + v_{0i}\right), 0\right]^T - \frac{1}{2} \tilde{W}_1^T \sigma''\left(V^T x_1 + v_0\right) V^T \left[\epsilon_2^2(u_d, \alpha), 0\right]^T [1\ 0]\, V$$
$$-W_1^T \sigma''\left(V^T x_1 + v_0\right) \left[\tilde{W}_2^T \sigma_i\left(V_i^T x_2 + v_{0i}\right), 0\right] V V^T \left[W_2^T \sigma_i\left(V_i^T x_2 + v_{0i}\right), 0\right]^T$$
$$-W_1^T \sigma''\left(V^T x_1 + v_0\right) V^T [1, 0]^T \left[\tilde{W}_2^T \sigma_i\left(V_i^T x_2 + v_{0i}\right) \epsilon_2(u_d, \alpha), 0\right] V + \frac{1}{2} W_1^T \sigma''$$
$$\cdot\left(V_i^T x_2 + v_{0i}\right) V^T \left[\hat{u}_{dNN}, 0\right]^T \left[W_2^T \sigma_i\left(V_i^T x_2 + v_{0i}\right), 0\right] V - W_1^T R_2\left(\tilde{W}_2, u_d, \alpha\right).$$

The importance of (6.12) is that it replaces higher order terms with linear terms of NN weight approximation errors $\tilde{W}_1$, $\tilde{W}_2$. This expression of $u - u_d$ is desirable for the adaptive compensation design. The expressions (6.11) and (6.12) can be reduced to the first-order approximation by using the first Taylor polynomial instead of second Taylor polynomial used in this paper. The reminder term is that case is of degree 2.

6.1.2 Approximation Error Analysis

The choice of W_1, V, v_0 and v_{0i} is based on the space $x = [v, u_d, \alpha]^T$. The term $R_2\left(\tilde{W}_2, u_d, \alpha\right)$ is bounded as

$$\|R_2(\cdot)\|_F \le \frac{1}{6} \|\sigma'''(\cdot)\| \|V\|_F^3 \left(\|\tilde{W}_2\|_F^3 \|\sigma(\cdot)\|^3 + \|\epsilon_2(u_d, \alpha)\|^3\right.$$
$$\left. 3\|\tilde{W}_2\|_F^2 \|\sigma(\cdot)\|^2 \|\epsilon_2(u_d, \alpha)\| + 3\|\tilde{W}_2\|_F \|\sigma(\cdot)\| \|\epsilon_2(u_d, \alpha)\|^2\right), \quad (6.14)$$

where $\sigma'''(x)$ is the third derivative of $\sigma(x)$ with respect to x: $\sigma''' = \frac{d^3\sigma(x)}{dx^3}$. Using (6.14), we get

$$|\epsilon| \le \|W_1\|_F \|\sigma'(\cdot)\| \|V\|_F \|\epsilon_2(u_d, \alpha)\| + \|\tilde{W}_1\|_F \|\sigma'(\cdot)\| \|V\|_F \|\tilde{W}_2\|_F \|\sigma_i(\cdot)\|$$
$$+\frac{1}{2} \|\sigma''(\cdot)\| \|V\|_F^2 \left(\|W_1\|_F \|\epsilon_2(u_d, \alpha)\|^2 + \|\tilde{W}_1\|_F \|\hat{W}_2\|_F \|\sigma_i(\cdot)\|^2 \|W_2\|_F\right)$$
$$+\|W_1\|_F \|\sigma''(\cdot)\| \|V\|_F^2 \|\tilde{W}_2\|_F \left(\|\sigma(\cdot)_i\| \|\epsilon_2(u_d, \alpha)\| + \|W_2\|_F \|\sigma_i(\cdot)\|^2\right)$$
$$+\frac{1}{6} \|\sigma'''(\cdot)\| \|V\|_F^3 \left(\|\tilde{W}_2\|_F^3 \|\sigma(\cdot)\|^3 + \|\epsilon_2(u_d, \alpha)\|^3\right.$$
$$\left. 3\|\tilde{W}_2\|_F^2 \|\sigma(\cdot)\|^2 \|\epsilon_2(u_d, \alpha)\| + 3\|\tilde{W}_2\|_F \|\sigma(\cdot)\| \|\epsilon_2(u_d, \alpha)\|^2\right)$$
$$= \bar{\delta}_1 \|\tilde{W}_2\|_F^3 + \bar{\delta}_2 \|\tilde{W}_2\|_F^2 + \bar{\delta}_3 \|\tilde{W}_2\|_F + \bar{\delta}_4 \|\tilde{W}_2\|_F \|\tilde{W}_1\|_F + \bar{\delta}_5 \|\tilde{W}_1\|_F + \bar{\delta}_6, \quad (6.15)$$

where $\bar{\delta}_i$, $i = 1, \ldots, 6$, are constant but unknown bounds. For assumptions on the bounds as mentioned earlier, and

$$\|\tilde{W}_1\|_F \leq W_{1M} + \|\hat{W}_1\|_F, \quad \|\tilde{W}_2\|_F \leq W_{2M} + \|\hat{W}_2\|_F, \tag{6.16}$$

we get

$$|\epsilon| \leq \delta_1\|\hat{W}_2\|_F^3 + \delta_2\|\hat{W}_2\|_F^2 + \delta_3\|\hat{W}_2\|_F + \delta_4\|\hat{W}_2\|_F\|\hat{W}_1\|_F + \delta_5\|\hat{W}_1\|_F + \delta_6 = \xi^T\Phi, \tag{6.17}$$

where $\xi = [\delta_1, \ldots, \delta_6]^T$ are bounded, unknown and adaptively updated, and $\Phi = \left[\|\hat{W}_2\|_F^3, \|\hat{W}_2\|_F^2, \|\hat{W}_2\|_F, \|\hat{W}_2\|_F, \|\hat{W}_1\|_F, \|\hat{W}_1\|_F, 1\right]^T$ is a known vector. The vector ξ is bounded because $\epsilon_2(u_d, \alpha)$ is bounded for bounded α. We define an estimator for the bound of the model mismatch error $\xi^T\Phi$, as

$$\hat{\epsilon} = \hat{\xi}^T\Phi, \tag{6.18}$$

where $\hat{\xi} = \left[\hat{\delta}_1, \ldots, \hat{\delta}_6\right]^T$ is estimated using adaptive laws which rely on projection functions.

6.2 A Benchmark Pitch-Plane Control System

Figure 6.1 presents a twin NN based uncertainty compensator framework. The first approximator estimates actuator nonlinearities, while the second serves as the inverse function compensator. The thrust T is an external (not from the actuators) input.

A nonlinear control signal similar to the one developed for low AoA [7] can now be designed using the control error (6.12). Of course, the presence of the model mismatch error ϵ in (6.12) requires a different adaptation scheme and would provide only asymptotic tracking within a bounded region.

An explicit adaptive inverse formulation for a SJA model at varying AoA is not possible even approximately with simply an approximated actuator model because

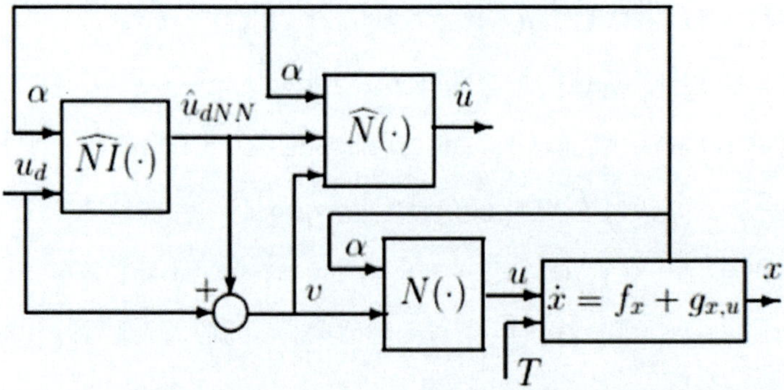

Fig. 6.1 AIC scheme with twin NNs

the structure of the model and the corresponding parameters are significantly different at different AoA. Therefore, we wish to formulate an AIC framework applicable to SJAs for aircraft operating at a wide range of AoA. We utilize the adaptive inverse scheme presented in the previous section and apply it to aircraft flight control system operating at a wide range of conditions, including both low and high AoA. The jets at the trailing edge are activated while operating at low AoA, and the leading edge actuator array is activated while operating at high AoA [8]. At intermediate AoA, one possible approach is to optimally select some jets from both the leading and trailing edges of the wing.

The states of the pitch-plane aircraft are

$$x = \begin{bmatrix} V & \gamma & \alpha & q \end{bmatrix}, \tag{6.19}$$

where V is the relative aircraft velocity, γ is the flight-path angle, q is the pitch angle, and α is the angle of attack.

The pitch-plane dynamics are

$$\dot{V} = \frac{T}{m}\,\mathrm{cs}\alpha - g\,\mathrm{sn}\,\gamma, \tag{6.20}$$

$$\dot{\gamma} = \frac{1}{mV}\left(u + T\,\mathrm{sn}\,\alpha - mg\,\mathrm{cs}\gamma\right), \tag{6.21}$$

$$\dot{\alpha} = q - \frac{1}{mV}\left(u + T\,\mathrm{sn}\,\alpha - mg\,\mathrm{cs}\gamma\right), \tag{6.22}$$

$$\dot{q} = \frac{I_{zz} - I_{xx}}{I_{yy}}\tau u, \tag{6.23}$$

where cs and sn represent cosine and sine functions respectively, u is the net lift force, m is the aircraft mass, τu is the pitch moment, T is the thrust, and inertias in the body axes are I_{ij}, $i = x, y, z$; $j = x, y, z$.

In the wing dynamics, the control inputs are the thrust signal T and the lift force $u(t)$ which represent the output of the SJAs. For the feedback control design, we assume that

(A2) the velocity V does not approach 0, the angle γ does not approach $\pm 90°$, and $-24° \leq \alpha \leq 24°$, so that the nonlinear aircraft dynamics avoid singularities.

The velocity dynamics (6.20) do not have an input from the actuators with uncertainties. Therefore, the input signal T is a good candidate to be generated from this subsystem and utilized for control of other subsystems. For velocity control, the desired thrust T can be generated as shown in Chap. 3, and for velocity error $\tilde{V} = V - V_r$, and constant gain $K_V > 0$, the velocity error dynamics is given by

$$\dot{\tilde{V}} = -K_V \tilde{V}. \tag{6.24}$$

With thrust T and a known f_γ, the flight path angle dynamics (6.21) can be given by

$$\dot{\gamma} = f_\gamma + \frac{1}{mV}\left(u + T\,\mathrm{sn}\,\alpha\right). \tag{6.25}$$

From (6.25), we calculate

$$\alpha_r = \arcsin\left(\frac{mVF_\gamma - u_d(t) - \hat{\xi}^T\Phi}{T}\right), \tag{6.26}$$

where $F_\gamma = \dot{\gamma}_r - f_\gamma - K_\gamma\tilde{\gamma}$, γ_r is from a reference system, $K_\gamma > 0$ is a constant gain, and $\tilde{\gamma} = \gamma - \gamma_r$ is flight path error. The solution of α_r is valid in the region

$$\left|mVF_\gamma - u_d - \hat{\xi}^T\Phi\right| < \delta|T|, \tag{6.27}$$

where δ is chosen by the allowable values of $\alpha_r : \alpha_r \in (-25°, 25°)\,, \delta = 0.1736$. Using (6.25)–(6.27), the dynamics of $\tilde{\gamma}$ are

$$\dot{\tilde{\gamma}} = -K_\gamma\tilde{\gamma} + \frac{1}{mV}\left(u - u_d - \hat{\xi}^T\Phi + T\left(\mathrm{sn}\,\alpha - \ \mathrm{sn}\,\alpha_r\right)\right). \tag{6.28}$$

For small errors $\tilde{\alpha} = \alpha - \alpha_r$, we get

$$\dot{\tilde{\gamma}} = -K_\gamma\tilde{\gamma} + \frac{1}{mV}\left(u - u_d - \hat{\xi}^T\Phi + T\tilde{\alpha}\,\mathrm{cs}\alpha_r\right) + \frac{T}{mV}O(\tilde{\alpha}). \tag{6.29}$$

Note that (6.29) is preferred over (6.28) because of the error terms being linear and hence handled with ease during Lyapunov analysis.

With thrust T, the angle of attack dynamics (6.22) is given by

$$\dot{\alpha} = f_\alpha - \frac{1}{mV}u + q, \tag{6.30}$$

where f_α is a known function. For angle of attack control, we generate

$$q_r = -f_\alpha + \dot{\alpha}_r - K_\alpha\tilde{\alpha} + \frac{1}{mV}\left(u_d + \hat{\xi}^T\Phi + T\tilde{\gamma}\,\mathrm{cs}\alpha_r\right), \tag{6.31}$$

where α_r is a reference signal, $K_\alpha > K_\gamma > 0$ are constant gains, $\tilde{\alpha} = \alpha - \alpha_r$ is the angle of attack error. The signal q_r would be useful as a reference signal for the pitch rate control. Note that the $\hat{\xi}^T\Phi$ and $T\tilde{\gamma}\,\mathrm{cs}\alpha_r$ terms are introduced strategically for lyapunov stability.

Using (6.30), (6.31), and (6.12), considering $\tilde{q}$ as the body axis rate error, the angle of attack error dynamics $\tilde{\alpha}$ is

$$\dot{\tilde{\alpha}} = -K_\alpha \tilde{\alpha} + \tilde{q} - \frac{1}{mV}\left(u - u_d - \hat{\xi}^T \Phi\right). \tag{6.32}$$

The control law of the entire system is given as

$$u_d = \frac{1}{\tau}\left(\frac{I_{yy}}{I_{zz} - I_{xx}}\right)(-K_q \tilde{q} - \tilde{\alpha} + \dot{q}_r) - \hat{\xi}^T \Phi, \tag{6.33}$$

where $K_q > K_\alpha > 0$, so that q_r (6.31) is attainable. The first term in (6.33) accommodates a negative feedback of the measured tracking error $\tilde{q}$. Using the pitch plane dynamics (6.23) and the control law (6.33), the pitch plane error dynamics are

$$\dot{\tilde{q}} = -K_q \tilde{q} - \tilde{\alpha} + \left(\frac{I_{zz} - I_{xx}}{I_{yy}}\right)\tau\left(u - u_d - \hat{\xi}^T \Phi\right), \tag{6.34}$$

where $\tilde{q} = q - q_r$ and $u - u_d$ is given in (6.12).

The control objective is to design of parameter estimates $\hat{W}_1$ and $\hat{W}_2$ for closed-loop semi-global stability with $\tilde{x}$ eventually converging to a compact set to be defined subsequently.

6.3 Parameter Adaptation and Stability Analysis

We formulate the adaptive update laws to determine the estimation of the parameters for achieving the control objective, as follows:

$$\begin{aligned}
\dot{\hat{W}}_1(t) &= g_1(t) + h_1(t), \\
\dot{\hat{W}}_2(t) &= g_2(t) + h_2(t), \\
\dot{\hat{\xi}}(t) &= \Gamma_3 \Phi |e(t)| + h_\xi(t),
\end{aligned} \tag{6.35}$$

where $g_1(t)$ and $g_2(t)$ are expressed as

$$\begin{aligned}
g_1(t) &= \Gamma_1 e(t) \Psi [\hat{u}_{dNN}, 0]^T, \\
g_2(t) &= -\Gamma_2 e(t) \left[\sigma_i\left(V_i^T x_2 + v_{0i}\right), 0\right]^T \Psi^T \hat{W}_1,
\end{aligned} \tag{6.36}$$

with Ψ given by (6.13) and where $\Gamma_i, i = 1, 2, 3$ are the adaptation gains such that $\Gamma_i = \Gamma_i^T > 0$ and

$$e(t) = \frac{\tilde{\gamma} - \tilde{\alpha}}{mV} + \frac{I_{zz} - I_{xx}}{I_{yy}} \tau \tilde{q}, \tag{6.37}$$

and h_1, h_2, h_ξ are the parameter projections ensuring parameter estimations are constrained in a desired region.

For parameter projection we denote W_{1_j}, $\hat{W}_{1_j}(t)$, W_{2_j}, $\hat{W}_{2_j}(t)$, $h_{1_j}(t), h_{2_j}(t)$ as the jth components of W_1, $\hat{W}_1(t)$, W_2, $\hat{W}_2(t)$, $h_1(t), h_2(t)$, respectively, for $j = 1, 2, \ldots, n_1$ and $j = 1, 2, \ldots, n_2$ as the case may be. The convex region that contains the ideal target weights for the two NNs is specified as

$$W_{1_j} \in \left[W_{1_j}^a, W_{1_j}^b\right], j = 1, 2, \ldots, n_1, \quad W_{2_j} \in \left[W_{2_j}^a, W_{2_j}^b\right], j = 1, 2, \ldots, n_2. \tag{6.38}$$

With $\hat{W}_{1_j}(0) \in \left[W_{1_j}^a, W_{1_j}^b\right]$, $\hat{W}_{2_j}(0) \in \left[W_{2_j}^a, W_{2_j}^b\right]$, the projection functions are

$$h_{1_j}(t) = \begin{cases} 0 & \text{if } \hat{W}_{1_j}(t) \in \left(W_{1_j}^a, W_{1_j}^b\right), \quad \text{or} \\ & \text{if } \hat{W}_{1_j}(t) = W_{1_j}^a \text{ and } g_{1_j}(t) \geq 0, \\ & \text{if } \hat{W}_{1_j}(t) = W_{1_j}^b \text{ and } g_{1_j}(t) \leq 0, \\ -g_{1_j}(t) & \text{otherwise}, \end{cases} \tag{6.39}$$

$$h_{c_j}(t) = \begin{cases} 0 & \text{if } \hat{W}_{2_j}(t) \in \left(W_{2_j}^a, W_{2_j}^b\right), \quad \text{or} \\ & \text{if } \hat{W}_{2_j}(t) = W_{2_j}^a \text{ and } g_{2_j}(t) \geq 0, \\ & \text{if } \hat{W}_{2_j}(t) = W_{2_j}^b \text{ and } g_{2_j}(t) \leq 0, \\ -g_{2_j}(t) & \text{otherwise}, \end{cases} \tag{6.40}$$

$$h_{\xi_k}(t) = \begin{cases} 0 & \text{if } \hat{\delta}_k(t) \ \epsilon(\delta_k^b, \delta_k^d), \quad \text{or} \\ & \text{if } \hat{\delta}_k(t) = \delta_k^b \text{ and } \Gamma_3\Phi e(t) \leq 0, \\ & \text{if } \hat{\delta}_k(t) = \delta_k^d \text{ and } \Gamma_3\Phi e(t) \geq 0, \\ \Gamma_3\Phi e(t) & \text{otherwise}, \end{cases} \tag{6.41}$$

where $k = 1, \ldots, 4$, $h_\xi = [h_{\xi_1}, \ldots, h_{\xi_4}]^T$. It is required that the initial weights $\hat{W}_1(0)$ and $\hat{W}_2(0)$ cannot both be initialized at zero since in that case these weights would stay at zero forever, because of the coupled nature as seen from the adaptive laws (6.35) and (6.36). Next, we provide the stability analysis for the system under study.

Theorem 1 *Under assumptions (A1) and (A2), the error dynamics (6.24), (6.28), (6.32), and (6.34) from state feedback control law governed by (6.26), (6.31) and (6.33), with parameters updated from the adaptive laws (6.35) and (6.36) guarantee semi-global closed-loop stability such that all of the signals are bounded and* $\tilde{x} = \left[V \;\; \tilde{\gamma} \;\; \tilde{\alpha} \;\; q\right]^T$ *eventually converges to a compact set*

$$S(\delta^2) = \left\{ \tilde{x} \,\middle|\, \int_t^{t+T} \|\tilde{x}(\tau)\|_2^2 \leq \delta^2 T + c_0 \right\}, \tag{6.42}$$

for a small $\tilde{x}$ *is small in the mean squared sense, and constant* $c_0 > 0$.

Proof Let us consider a scalar continuous and positive definite function

$$V_L = \frac{1}{2}\tilde{x}^T\tilde{x} + \frac{1}{2}\tilde{W}_1^T\Gamma_1^{-1}\tilde{W}_1 + \frac{1}{2}\tilde{W}_2^T\Gamma_2^{-1}\tilde{W}_2 + \frac{1}{2}\tilde{\xi}^T\Gamma_3^{-1}\tilde{\xi}, \tag{6.43}$$

where $\tilde{\xi}(t) = \xi - \hat{\xi}(t)$. Using (6.24), (6.28), (6.32), (6.34) and (6.37), and differentiating V_L, we have

$$\dot{V}_L = -K_x\tilde{x} + e(t)\left(u - u_d - \hat{\xi}^T\Phi\right) + \Delta - \tilde{W}_1^T\Gamma_1^{-1}\dot{\hat{W}}_1 - \tilde{W}_2^T\Gamma_2^{-1}\dot{\hat{W}}_2 - \tilde{\xi}^T\Gamma_3^{-1}\dot{\hat{\xi}}, \tag{6.44}$$

where $\Delta = \frac{T\tilde{\gamma}}{mV}O(\tilde{\alpha}) \leq \eta\tilde{\gamma}$ for higher order $O(\tilde{\alpha})$ and a bound η which can be computed. Substituting the control error (6.12) and adaptive laws (6.35) in (6.44), we have

$$\begin{aligned}\dot{V}_L = & -K_x\tilde{x} + \Delta - \tilde{W}_1^T\Gamma_1^{-1}h_1(t) - \tilde{W}_2^T\Gamma_2^{-1}h_2(t) - \tilde{\xi}^T\Gamma_3^{-1}h_\xi(t) - e(t)\hat{\xi}^T\Phi \\ & -\tilde{W}_1 e(t)\Psi\left[\hat{u}_{dNN}, 0\right]^T + \tilde{W}_2 e(t)\left[\sigma_i\left(V_i^T x_2 + v_{0i}\right), 0\right]^T \Psi^T\hat{W}_1 - \tilde{\xi}^T\Phi|e(t)| \\ & +e(t)\left(\tilde{W}_1^T\Psi\left[\hat{u}_{dNN}, 0\right]^T - \hat{W}_1^T\Psi\left[\tilde{W}_2^T\sigma_i\left(V_i^T x_2 + v_{0i}\right), 0\right]^T + \epsilon\right).\end{aligned} \tag{6.45}$$

Since $|\epsilon| \leq \xi^T\Phi$,by cancelling certain terms appropriately, we get

$$\dot{V}_L = -K_x\tilde{x} - \tilde{W}_1^T\Gamma_1^{-1}h_1(t) - \tilde{W}_2^T\Gamma_2^{-1}h_2(t) - \tilde{\xi}^T\Gamma_3^{-1}h_\xi(t) + \eta\tilde{\gamma}. \tag{6.46}$$

With the parameter projections (6.39–6.41) and appropriate known bounds for the initial values for the parameter estimates, which maintain the parameters in the desired bounds, it follows that

$$\hat{W}_{1_j} \in \left[W_{1_j}^a, W_{1_j}^b\right], j = 1, \ldots, n_1, \ \ \hat{W}_{2_j} \in \left[W_{2_j}^a, W_{2_j}^b\right], j = 1, \ldots, n_2, \tag{6.47}$$

$$\tilde{W}_1^T(t)h_1(t) \geq 0, \ \ \tilde{W}_2^T(t)h_2(t) \geq 0, \ \ \tilde{\xi}^T h_\xi(t) \geq 0. \tag{6.48}$$

With this property, (6.46) reduces to

$$\dot{V}_L = -K_V\tilde{V} - K_\gamma\tilde{\gamma} - K_\alpha\tilde{\alpha} - K_q\tilde{q} + \eta\tilde{\gamma}. \tag{6.49}$$

Using the inequality

$$ab \leq \zeta^2x^2 + \frac{1}{4\zeta^2}y^2, \ \forall\zeta \neq 0, \tag{6.50}$$

for $\zeta^2 = \frac{K_\gamma}{2}, \ \ \sigma \geq |\eta|$, we get

$$\dot{V}_L = -\left[K_V \ \frac{K_\gamma}{2} \ K_\alpha \ K_q\right]\tilde{x}^2 + \frac{1}{2K_\gamma}\sigma^2. \tag{6.51}$$

Integrating both sides over the time interval $[t, t+T]$, we have

$$\int_t^{t+T} \dot{V}_L(\tau)d\tau \leq -\|K\| \int_t^{t+T} \|\tilde{x}(\tau)\|_2^2 d\tau + \frac{\sigma^2}{2K_\gamma}T, \tag{6.52}$$

where $K = \left[K_V \ \frac{K_\gamma}{2} \ K_\alpha \ K_q\right]$, and therefore, we have

$$\int_t^{t+T} \|\tilde{x}(\tau)\|_2^2 d\tau \leq \delta^2 T + c_0, \forall t \geq 0, \text{ and any } T \geq 0, \tag{6.53}$$

where c_0 and δ^2 are given as

$$c_0 = \sup_{t\geq 0} \frac{1}{\|K\|}\left(V_L(t) - V_L(t+T)\right), \ \delta^2 = \frac{\sigma^2}{2\|K\|K_\gamma}. \tag{6.54}$$

We have $\tilde{x}(\tau) \in S(\delta^2)$, where appropriate choice of the parameters ensures the suitable adjustment of δ and where $\tilde{W}_1, \tilde{W}_2$ are also uniformly bounded. The new adaptive inverse control scheme ensures closed-loop signal boundedness and good tracking performance in the presence of model mismatch $\epsilon(t)$. The accuracy of adaptive compensation is increased with usage of higher-order approximation.

∇

References

1. Liu, L., Yin, S., Gao, H., Alsaadi, F., Hayat, T.: Adaptive partial-state feedback control for stochastic high-order nonlinear systems with stochastic input-to-state stable inverse dynamics. Automatica **51**, 285–291 (2015). https://doi.org/10.1016/j.automatica.2014.10.094
2. Cybenko, G.: Approximations by superpositions of sigmoidal functions. Math. Control, Signals, Syst. **2**(4), 303–314 (1989). https://doi.org/10.1007/BF02551274
3. Hornik, K.: Approximation capabilities of multilayer feedforward networks. Neural Netw. **4**(2), 251–257 (1991). https://doi.org/10.1016/0893-6080(91)90009-t
4. Deb, D., Tao, G., Burkholder, J.: High-order design of adaptive inverses for signal-dependent actuator nonlinearities. In: 2008 American Control Conference (2008). https://doi.org/10.1109/acc.2008.4587029
5. Lewis, F., Campos, J., Selmic, R.: Neuro-Fuzzy Control of Industrial Systems with Actuator Nonlinearities. Society for Industrial and Applied Mathematics. Philadelphia, PA (2002)
6. Selmic, R., Lewis, F.: Deadzone compensation in motion control systems using neural networks. IEEE Trans. Autom. Control **45**(4), 602–613 (2000). https://doi.org/10.1109/9.847098
7. Deb, D., Tao, G., Burkholder, J., Smith, D.: Adaptive compensation control of synthetic jet actuator arrays for airfoil virtual shaping. J. Aircr. **44**(2), 616–626 (2007). https://doi.org/10.2514/1.24910

8. Deb, D., Tao, G., Burkholder, J.: An adaptive inverse compensation scheme for signal-dependent actuator nonlinearities. In: 2007 46th IEEE Conference on Decision and Control (2007). https://doi.org/10.1109/cdc.2007.4434743
9. Stevens, B.L., Lewis, F.L.: Aircraft Control and Simulation. Wiley, New York, NY (1992)
10. Bryson Jr., A.E.: Control of Spacecraft and Aircraft. Princeton University Press, Princeton, NJ (1994)
11. Tao, G.: Adaptive Control Design and Analysis. Wiley (2003)
12. Liu, Y., Deb, D., Tao, G.: Modeling and multivariable adaptive control of aircraft with synthetic jet actuators. In: 2008 7th World Congress on Intelligent Control and Automation (2008). https://doi.org/10.1109/wcica.2008.4593263
13. Deb, D., Sonowal, S.: Synthetic jet actuator based adaptive neural network control of non-linear fixed pitch wind turbine blades. In: 2013 IEEE International Conference on Control Applications (CCA) (2013). https://doi.org/10.1109/cca.2013.6662759
14. Nath, A., Deb, D., Dey, R.: An augmented subcutaneous type 1 diabetic patient modelling and design of adaptive glucose control. J. Process. Control **86**, 94–105 (2020). https://doi.org/10.1016/j.jprocont.2019.08.010
15. Nath, A., Deb, D., Dey, R.: Robust observer based adaptive control of blood glucose in diabetic patients. Int. J. Control **1–0** (2020). https://doi.org/10.1080/00207179.2020.1750705
16. Deb, D., Tao, G., Burkholder, J., Smith, D.: An adaptive inverse control scheme for a synthetic jet actuator model. In: Proceedings of the 2005, American Control Conference (2005). https://doi.org/10.1109/acc.2005.1470367
17. Deb, D., Tao, G., Burkholder, J., Smith, D.: Adaptive synthetic jet actuator compensation for a nonlinear tailless aircraft model at low angles of attack. In: 2006 American Control Conference (2006). https://doi.org/10.1109/acc.2006.1655376
18. Deb, D., Tao, G., Burkholder, J., Smith, D.: An Adaptive Inverse Control Scheme for Synthetic Jet Actuator Arrays. In: Infotech@Aerospace (2005). https://doi.org/10.2514/6.2005-7170
19. Deb, D., Tao, G., Burkholder, J., Smith, D.: Adaptive synthetic jet actuator compensation for a nonlinear aircraft model at low angles of attack. IEEE Trans. Control Syst. Technol. **16**(5), 983–995 (2008). https://doi.org/10.1109/tcst.2007.912124

Chapter 7
Adaptive Synthetic Jet Actuation for Aircraft Control

The objective of this chapter is to develop a unified actuator nonlinearity compensation scheme for an aircraft augmented with SJs operating at a wide range of conditions, including both low and high AoA. The motivation for such a design is the intrinsically complicated nonlinear characteristic of a SJ when used for an aircraft operating in a wide range of AoA.

7.1 Wide-Range Angles of Attack Model

Because of complexity and uncertainty of the actuator model at different AoA, the construction of the inverse description is a challenge task. Moreover, an approximation based cancellation without considering this variation caused by a change in AoA, introduces compensation error and new challenges in closed-loop stability.

Low AoA Model: At a typical low AoA ($\alpha < 10°$), the parametrized model [1] for a non-linearity $N(\cdot)$ is given by

$$u(t) = N(v(t)) = \phi_b^* - \frac{\phi_a^*}{v(t)}, \tag{7.1}$$

for some parameters ϕ_a^*, ϕ_b^*, where $u = C_L$ is the lift coefficient (proportional to the lift force) and $v = A_{pp}^2$, with A_{pp}^2 being the input peak-to-peak voltage amplitude. This model is given by

$$u = N(v) = \phi^* \omega(t), \tag{7.2}$$

for $\phi^* = [\phi_a^*, \phi_b^*]^T$ and $\omega(t) = [-1/v(t), 1]$ which leads to the desired adaptive inverse characteristic

D. Deb et al., *Adaptive Compensation of Nonlinear Actuators for Flight Control Applications*, Studies in Systems, Decision and Control 386,
https://doi.org/10.1007/978-981-16-4161-9_7

$$v(t) = \widehat{NI}(u_d(t); \phi) = \frac{\phi_a(t)}{\phi_b(t) - u_d(t)}, \quad \phi = [\phi_a, \phi_b]^T, \tag{7.3}$$

where ϕ_1 and ϕ_2 are adaptive estimates of ϕ_1^*, ϕ_2^* and where u_d is a desired feedback control signal. For this inverse function, it is verified that for $\phi = \phi^*$, the nonlinearity $N(\cdot)$ is perfectly compensated, that is, $u = N(v) = N\left(\widehat{NI}\left(u_d, \phi^*\right)\right) = u_d$.

High AoA Model: At a typical high AoA ($24° > |\alpha| > 22°$), the input-output relationship is

$$u = \phi_2^{h*} + \phi_1^{h*} sin^2\left(\phi_3^{h*} v\right), \tag{7.4}$$

for some parameters ϕ_i^{h*}, $i = 1, 2, 3$. From (7.4), we can see that the jet characteristic at high AoA is nonlinearly parametrized for which the construction of an adaptive inverse is an open task of research. It is crucial to develop a suitable parametrization of the SJ characteristic for adaptive compensation.

For this model, unless ϕ_3^{h*} is known, a linearly parametrized form, that is, $u = N(v) = \phi^{*T}\omega(t)$ for some parameter vector ϕ^* and some known signal vector $\omega(t)$, is not available, but such a linear form is crucial for the design of an adaptive law to update an adaptive estimate of ϕ^*. We have to employ some approximation scheme to obtain an approximate linear parametrization of the jet model $N(\cdot)$.

From a comparison of (7.1) and (7.4), it is clear that a change in the AoA influences the SJ characteristics. Moreover, for a chosen linear parametrization $u = N(v) = \phi_a^*\omega_1(t)$ of the jet nonlinearity (where ω_1 depends on the signal v), the construction of its adaptive inverse $v = \widehat{NI}(u_d)$ for a designed state feedback signal $u_d(t)$ is based on the solution of the equation $u_d(t) = \phi_1^T(t)\omega(t)$ for v(t) in order to obtain an adaptive estimate $\phi(t)$ of ϕ^* (recall that $\omega(t)$ contains $v(t)$).

Unified AoA Model: Let $f_1(v)$ and $f_2(v)$ be jet nonlinearity functions at a low AoA ($\alpha_1 = 3°$) and high AoA ($\alpha_2 = 24°$) respectively. We build a nonlinearity profile given as

$$f(v, \alpha) = a(\alpha) f_1(v) + b(\alpha) f_2(v), \tag{7.5}$$

where $a(\alpha)$ and $b(\alpha)$ satisfy

$$a(\alpha_1) = 1, \;\; a(\alpha_2) = 0, \;\; b(\alpha_1) = 0, \;\; b(\alpha_2) = 1. \tag{7.6}$$

There are many such functions $a(\alpha)$ and $b(\alpha)$. The simplest ones (when we know the data at two different AoA) are linear functions. Using (7.6), we have

$$a(\alpha) = -0.0479\alpha + 1.1429, \;\; b(\alpha) = 0.0479\alpha - 0.1429.$$

The nonlinearity functions at $\alpha = 3°$ and $\alpha = 24°$ are given by $f_1(v) = 15 - \frac{33:33}{v}$ [2, 3] and $f_2(v) = 20 + 5\,\mathrm{sn}^2\left(\pi \frac{v}{32}\right)$ respectively. As more information becomes available (through further wind tunnel tests) about the actuator characteristics at different AoA, these functions can be made more precise. The function $f(v; \alpha)$ is valid for $v \in (0, 16)$ and $\alpha \in [1°, 24°]$.

An explicit solution v of such an equation $u_d = \phi_1^T \omega_1$, is difficult to obtain if $u_d = \phi_1^T \omega_1$ is an approximation of $u = N(v)$ and the order of approximation is high. If, however, the "physical" parameters in $N(\cdot)$ like ϕ_1^{h*}, ϕ_2^{h*} and ϕ_3^{h*} in (7.4) are estimated by an adaptive law (if possible), one can construct an adaptive inverse

$$v = \widehat{NI}\left(u_d\right) = N^{-1}\left(u_d; \phi_1^h, \phi_2^h, \phi_3^h\right) \tag{7.7}$$

for the nonlinearity $u = N\left(v; \phi_1^{h*}, \phi_2^{h*}, \phi_3^{h*}\right)$, where $\phi_1^h, \phi_2^h, \phi_3^h$ are the adaptive estimates of $\phi_1^{h*}, \phi_2^{h*}, \phi_3^{h*}$ and where $N^{-1}(\cdot)$ is the inverse of $N(\cdot)$. An explicit adaptive inverse formulation at varying AoA is not possible even approximately with simply an approximated actuator model because the structure of the model and the corresponding parameters are significantly different at different AoA. Therefore, the approach of adaptive inverse formulation has to be revisited.

Such an inverse function appropriate for the entire range of AoA can be designed either using switching mechanisms based on the changes of AoA over certain pre-specified sub-intervals of AoA or using a functional approximation-based parametrization scheme. It is the task of this chapter to develop such an approximation based adaptive scheme.

7.2 Adaptive Inverse Design of Actuator Nonlinearities

Apart from the AoA and the input voltage signal, the number of active jets and location of the jets are critical implicit attributes of the actuator model characteristics. Next, we explore the adaptive formulation for a SJA model

$$u = N\left(\phi^*, v, \alpha\right), \tag{7.8}$$

where u is the actual signal (output) from the SJAs which is not measurable, v is the actuator input vector, α is the AoA which contributes to the actuator characteristics, and ϕ^* is a vector containing uncertain parameters. This parameter vector varies widely with AoA and the nonlinearly parametrized model is difficult to construct and compensate for.

Such a model with input voltage amplitude and AoA as the actuator input signals encapsulates the affect of actuator failures or affect of the location of the actuators. That is, the adaptive compensation control scheme to be developed, is robust to actuator failures. It is well known that NNs, fuzzy approximators, and Taylor series approximators have an *universal approximation* property; that is, for any continuous function $f(v, \alpha)$, and for a compact set S such that $v, \alpha \in S \subset R^n$, from the generic development in Chap. 5 (and [4]) with $\bar{x} = \alpha$, the nonlinearity model for a SJA is

$$u = N(v, \alpha) \triangleq f(v, \alpha) = \phi_1^{*T} \omega_1(v, \alpha) + \epsilon_1(v, \alpha), \tag{7.9}$$

where ϕ_1^* is an unknown parameter vector expressed as

$$\phi_1^* = \left[f(a,b), f^{(0,1)}(a,b), f^{(1,0)}(a,b), f^{(1,1)}(a,b), \ldots, f^{(M_1-1,R_1)}(a,b),\right.$$
$$\left. f^{(M_1,R_1-1)}(a,b), f^{(M_1,R_1)}(a,b)\right]^T \in R^{(M_1+1)(R_1+1)}, \tag{7.10}$$

and $\epsilon_1(v,\alpha)$ is the modeling error, and $\omega_1(v,\alpha)$ is a known vector function given by

$$\omega_1 = \left[1, \frac{\alpha-b}{1!}, \frac{v-a}{1!}, \frac{(v-a)(\alpha-b)}{1!1!}, \ldots, \frac{(v-a)^{M_1-1}(\alpha-b)^{R_1}}{(M_1-1)!R_1!},\right.$$
$$\left. \frac{(v-a)^{M_1}(\alpha-b)^{R_1-1}}{M_1!(R_1-1)!}, \frac{(v-a)^{M_1}(\alpha-b)^{R_1}}{M_1!R_1!}\right]^T, \tag{7.11}$$

with suitable real scalars a, b and a_1, a_2 such that the interval $|v-a| < a_1$ covers the range of $v \in (0, 16)$, and the interval $|\alpha - b| < a_2$ covers the range of $\alpha \in (-25°, 25°)$. The Taylor series approximation is

$$f(v,\alpha) = \sum_{m=0}^{M_1}\sum_{r=0}^{R_1} \frac{f^{(m,r)}(a,b)}{m!r!}(v-a)^m(\alpha-b)^r, \tag{7.12}$$

where $f^{(m,r)}(a,b) = \frac{\partial^m}{\partial v^m}\frac{\partial^r}{\partial \alpha^r} f(v,\alpha)|_{a,b}$ are smooth nonlinear and invertible functions, but unknown so that compensation is difficult. For a parametrized nonlinearity function $\bar{u} = \phi_1^{*T}\omega_1(v,\alpha)$, an estimate $\hat{u}$ can be formulated as

$$\hat{u} = \hat{N}(v,\alpha) = \hat{\phi}_1^T\omega_1(v,\alpha). \tag{7.13}$$

For the inverse characteristic such that $u_d = f\left(f^{-1}(u_d,\alpha),\alpha\right)$, we use another Taylor series approximator to estimate the function

$$u_{dT}(u_d,\alpha) = \widehat{NI}(u_d,\alpha) \stackrel{\triangle}{=} NI(u_d,\alpha) - u_d = f^{-1}(u_d,\alpha) - u_d, \tag{7.14}$$

where u_d is the desired state feedback signal, and u_{dT} is a modified nonlinearity inverse function given by

$$u_{dT}(u_d,\alpha) = \widehat{NI}(u_d,\alpha) \stackrel{\triangle}{=} g(u_d,\alpha) = \phi_2^{*T}\omega_2(u_d,\alpha) + \epsilon_2(u_d,\alpha), \tag{7.15}$$

where ϕ_2^* is an unknown parameter vector expressed as

$$\phi_2^* = \left[g(c,b), g^{(0,1)}(c,b), g^{(1,0)}(c,b), g^{(1,1)}(c,b), \ldots, g^{(M_2-1,R_2)}(c,b),\right.$$
$$\left. g^{(M_2,R_2-1)}(c,b), g^{(M_2,R_2)}(c,b)\right]^T \in R^{(M_2+1)(R_2+1)}, \tag{7.16}$$

and $\epsilon_2(u_d,\alpha)$ is the function modeling error, and $\omega_2(u_d,\alpha)$ is a known vector function

$$\omega_2 = \left[1, \frac{\alpha - b}{1!}, \frac{u_d - c}{1!}, \frac{(u_d - c)(\alpha - b)}{1!1!}, \ldots, \frac{(u_d - c)^{M_2 - 1}(\alpha - b)^{R_2}}{(M_2 - 1)!R_2!}, \right.$$
$$\left. \frac{(u_d - c)^{M_2}(\alpha - b)^{R_2 - 1}}{M_2!(R_2 - 1)!}, \frac{(u_d - c)^{M_2}(\alpha - b)^{R_2}}{M_2!R_2!}\right]^T, \tag{7.17}$$

and real scalars c, b and r_3, r_2 such that the interval $|u_d - c| < r_3$ covers the range of u_d generated from state feedback, and the interval $|\alpha - b| < r_2$ covers the range of $\alpha \in (-24°, 24°)$. The Taylor series approximation is given by

$$g(u_d, \alpha) = \sum_{m=0}^{M_2} \sum_{r=0}^{R_2} \frac{g^{(m,r)}(c, b)}{m!r!}(u_d - c)^m (\alpha - b)^r, \tag{7.18}$$

where $g^{(m,r)}(c, b) = \frac{\partial^m}{\partial v^m} \frac{\partial^r}{\partial \alpha^r} g(u_d, \alpha)|_{c,b}$ are smooth nonlinear functions.

For a parametrized nonlinearity function $\bar{u}_{dT}(u_d, \alpha) = \phi_2^{*T} \omega_2(u_d, \alpha)$, an estimate $\hat{u}_{dT}(u_d, \alpha)$ can be expressed as

$$\hat{u}_{dT}(u_d, \alpha) = NI(u_d, \alpha) = \hat{\phi}_2^T \omega_2(u_d, \alpha). \tag{7.19}$$

Using (7.14) and $\tilde{u}_{dT} = u_{dT} - \hat{u}_{dT}$, we get

$$u_d = f(u_d + \hat{u}_{dT} + \tilde{u}_{dT}, \alpha). \tag{7.20}$$

From Taylor series expansion taken about $u_{dT} = \hat{u}_{dT}$, and using the applied control

$$v = u_d + \hat{u}_{dT}, \tag{7.21}$$

we get the desired signal u_d given by

$$u_d = f(v, \alpha) + \frac{\partial f(v, \alpha)}{\partial v} \tilde{u}_{dT} + \frac{\partial f^2(v, \alpha)}{\partial v^2} \tilde{u}_{dT}^2 + O\left(\frac{\partial f^i(v, \alpha)}{\partial v^i}, \tilde{u}_{dT}\right), \tag{7.22}$$

where $i = 3, 4, \ldots,$ and $O(\cdot)$ represents the sum of the higher order arguments.

The estimate on the remainder of the second Taylor polynomial, $O\left(\frac{\partial f^i(v,\alpha)}{\partial v^i}, \tilde{u}_{dT}\right)$ describes how far the approximation is from the true function. The remainder term is only degree 3 and satisfies the inequality

$$\left\| O\left(\frac{\partial f^i(v, \alpha)}{\partial v^i}, \tilde{u}_{dT}\right) \right\| \leq \frac{1}{6} \left\| \phi^{*T} \frac{\partial^3 \omega_1(v, \alpha)}{\partial v^3} + \frac{\partial^3 \epsilon_1(v, \alpha)}{\partial v^3} \right\| \left\| \tilde{\phi}_2^T \omega_2(u_d, \alpha) + \epsilon_2(u_d, \alpha) \right\|^3$$
$$\leq a_1 \left\| \hat{\phi}_2 \right\|^3 + a_7 \left\| \hat{\phi}_2 \right\|^2 + a_8 \left\| \hat{\phi}_2 \right\| + a_9, \tag{7.23}$$

where the second inequality is obtained using the following assumption and where $\left\|\tilde{\phi}_2\right\| \leq \phi_2 + \left\|\hat{\phi}_2\right\|$, $a_i, i = 1, 7, 8, 9$ are bounded constants not assumed to be known.

(A1) the nominal parameters ϕ_1^* and ϕ_2^* are bounded such that $||\phi_1^*|| \leq \phi_A$ and $||\phi_2^*|| \leq \phi_2$ where ϕ_A and ϕ_2 are known, and the modelling errors $\epsilon_1(v, \bar{x})$ and $\epsilon_2(u_d, \bar{x})$ are bounded such that $||\epsilon_1|| \leq \epsilon_{N2}$.

7.2.1 Control Error Expression

Next, we describe the effectiveness of this coupled nonlinearity functional approximation by developing a control error expression.

Lemma 7.1 *Under assumption (A1), with the functional compensator described by (7.19) and (7.22) and the observer given by (7.13), the control error for the SJA nonlinearity compensation, is given by*

$$\begin{aligned} u - u_d = {} & \tilde{\phi}_1^T \left(\frac{\partial \omega_1(v, \alpha)}{\partial v} - \frac{\partial^2 \omega_1(v, \alpha)}{\partial v^2} \hat{\phi}_2^T \omega_2(u_d, \alpha) \right) \hat{\phi}_2^T \omega_2(u_d, \alpha) \\ & - \hat{\phi}_1^T \left(\frac{\partial \omega_1(v, \alpha)}{\partial v} - \frac{\partial^2 \omega_1(v, \alpha)}{\partial v^2} \tilde{\phi}_2^T \omega_2(u_d, \alpha) \right) \hat{\phi}_2^T \omega_2(u_d, \alpha) + \epsilon, \end{aligned} \tag{7.24}$$

where ϵ is a bounded model mismatch error.

Proof From (7.13) and (7.15), we have

$$\tilde{u}_{dT} = \tilde{\phi}_2^T \omega_2(u_d, \alpha) + \epsilon_2(u_d, \alpha), \tag{7.25}$$

where $\tilde{\phi}_2(t) = \phi_2^* - \hat{\phi}_2(t)$ is the parameter approximation error for the inverse nonlinearity function. Using (7.9) and (7.22), the control error is given by

$$u - u_d = -\frac{\partial f(v, \alpha)}{\partial v} \tilde{u}_{dT} - \frac{\partial^2 f(v, \alpha)}{\partial v^2} \tilde{u}_{dT}^2 - O\left(\frac{\partial f^i(v, \alpha)}{\partial v^i}, \tilde{u}_{dT}, \alpha \right). \tag{7.26}$$

Substituting (7.9) and (7.25) in (7.26), we get

$$\begin{aligned} u - u_d = {} & -\left(\phi_1^{*T} \frac{\partial \omega_1(v, \alpha)}{\partial v} + \frac{\partial \epsilon_1(v, \alpha)}{\partial v} \right) \left(\tilde{\phi}_2^T \omega_2(u_d, \alpha) + \epsilon_2(u_d, \alpha) \right) \\ & - \left(\phi_1^{*T} \frac{\partial^2 \omega_1(v, \alpha)}{\partial v^2} + \frac{\partial^2 \epsilon_1(v, \alpha)}{\partial v^2} \right) \left(\tilde{\phi}_2^T \omega_2(u_d, \alpha) + \epsilon_2(u_d, \alpha) \right)^2 \\ & - O\left(\frac{\partial f^i(v, \alpha)}{\partial v^i}, \tilde{u}_{dT}, \alpha \right). \end{aligned} \tag{7.27}$$

For parameter approximation error $\tilde{\phi}_1(t) = \phi_1^* - \hat{\phi}_1(t)$, we get

$$
\begin{aligned}
u - u_d =& \tilde{\phi}_1^T \frac{\partial \omega_1(v,\alpha)}{\partial v} \hat{\phi}_2^T \omega_2(u_d,\alpha) - \hat{\phi}_1^T \frac{\partial \omega_1(v,\alpha)}{\partial v} \tilde{\phi}_2^T \omega_2(u_d,\alpha) \\
& - \frac{1}{2} \tilde{\phi}_1^T \frac{\partial^2 \omega_1(v,\alpha)}{\partial v^2} \hat{\phi}_2^T \omega_2(u_d,\alpha) \hat{\phi}_2^T \omega_2(u_d,\alpha) \\
& + \frac{1}{2} \hat{\phi}_1^T \frac{\partial^2 \omega_1(v,\alpha)}{\partial v^2} \hat{\phi}_2^T \omega_2(u_d,\alpha) \tilde{\phi}_2^T \omega_2(u_d,\alpha) + \epsilon \\
=& \tilde{\phi}_1^T \left(\frac{\partial \omega_1(v,\alpha)}{\partial v} - \frac{\partial^2 \omega_1(v,\alpha)}{\partial v^2} \hat{\phi}_2^T \omega_2(u_d,\alpha) \right) \hat{\phi}_2^T \omega_2(u_d,\alpha) \\
& - \hat{\phi}_1^T \left(\frac{\partial \omega_1(v,\alpha)}{\partial v} - \frac{\partial^2 \omega_1(v,\alpha)}{\partial v^2} \tilde{\phi}_2^T \omega_2(u_d,\alpha) \right) \hat{\phi}_2^T \omega_2(u_d,\alpha) + \epsilon, \quad (7.28)
\end{aligned}
$$

where ϵ is the model mismatch error given as

$$
\begin{aligned}
\epsilon =& - \tilde{\phi}_1^T \frac{\partial \omega_1(v,\alpha)}{\partial v} \phi_2^{*T} \omega_2(u_d,\alpha) - \frac{\partial \epsilon_1(v,\alpha)}{\partial v} \tilde{\phi}_2^T \omega_2(u_d,\alpha) \\
& - \phi_1^{*T} \frac{\partial \omega_1(v,\alpha)}{\partial v} \epsilon_2(u_d,\alpha) - \frac{\partial \epsilon_1(v,\alpha)}{\partial v} \epsilon_2(u_d,\alpha) - O\left(\frac{\partial f^i(v,\alpha)}{\partial v^i}, \tilde{u}_{dT}, \alpha \right) \\
& + \frac{1}{2} \left(\tilde{\phi}_1^T \frac{\partial^2 \omega_1(v,\alpha)}{\partial v^2} \phi_2^{*T} + \frac{\partial^2 \epsilon_1(v,\alpha)}{\partial v^2} \tilde{\phi}_2^T \right) \omega_2(u_d,\alpha) \left(\tilde{\phi}_2^T \omega_2(u_d,\alpha) + \epsilon_2(u_d,\alpha) \right) \\
& - \frac{1}{2} \left(\tilde{\phi}_1^T \frac{\partial^2 \omega_1(v,\alpha)}{\partial v^2} \hat{\phi}_2^T - \hat{\phi}_1^T \frac{\partial^2 \omega_1(v,\alpha)}{\partial v^2} \tilde{\phi}_2^T \right) \omega_2(u_d,\alpha) \left(\phi_2^{*T} \omega_2(u_d,\alpha) + \epsilon_2(u_d,\alpha) \right) \\
& + \frac{1}{2} \left(\frac{\partial^2 \epsilon_1(v,\alpha)}{\partial v^2} + \phi_1^{*T} \frac{\partial^2 \omega_1(v,\alpha)}{\partial v^2} \right) \epsilon_2(u_d,\alpha) \left(\tilde{\phi}_2^T \omega_2(u_d,\alpha) + \epsilon_2(u_d,\alpha) \right). \quad (7.29)
\end{aligned}
$$

Using assumption (A1), we have $\|\tilde{\phi}_1\| \le \phi_A + \|\hat{\phi}_1\|$ and $\|\tilde{\phi}_2\| \le \phi_C + \|\hat{\phi}_2\|$ and (7.23), the norm of the model mismatch error $\epsilon(t)$ satisfies the following condition:

$$
\begin{aligned}
|\epsilon(t)| \le& \phi_C \left(\phi_A + \|\hat{\phi}_1\| \right) \left\| \frac{\partial \omega_1(v,\alpha)}{\partial v} \right\| \|\omega_2(u_d,\alpha)\| + a_9 + a_8 \left\| \hat{\phi}_2 \right\| + a_7 \left\| \hat{\phi}_2 \right\|^2 \\
& + a_1 \left\| \hat{\phi}_2 \right\|^3 + \left\| \frac{\partial \epsilon_1(v,\alpha)}{\partial v} \right\| \left(\phi_C + \|\hat{\phi}_2\| \right) \|\omega_2(u_d,\alpha)\| + \phi_A \left\| \frac{\partial \omega_1(v,\alpha)}{\partial v} \right\| \epsilon_{N2} \\
& + \frac{\phi_C}{2} \left(\phi_C + \|\hat{\phi}_2\| + \epsilon_{N2} \right) \left(\phi_A + \|\hat{\phi}_1\| \right) \left\| \frac{\partial^2 \omega_1(v,\alpha)}{\partial v^2} \right\| \|\omega_2(u_d,\alpha)\|^2 \\
& + \frac{\epsilon_{N2}}{2} \left(\left(\phi_C + \|\hat{\phi}_2\| \right) \|\omega_2(u_d,\alpha)\| + \epsilon_{N2} \right) \left(\left\| \frac{\partial \epsilon_1^2(v,\alpha)}{\partial v^2} \right\| + \phi_A \left\| \frac{\partial^2 \omega_1(v,\alpha)}{\partial v^2} \right\| \right) \\
& + \frac{\phi_C \|\hat{\phi}_1\| + \phi_A \|\hat{\phi}_2\|}{2} \left(\phi_C \|\omega_2(u_d,\alpha)\| + \epsilon_{N2} \right) \left\| \frac{\partial^2 \omega_1(v,\alpha)}{\partial v^2} \right\| \|\omega_2(u_d,\alpha)\| \\
& + \frac{1}{2} \left\| \frac{\partial \epsilon_1^2(v,\alpha)}{\partial v^2} \right\| \left(\phi_C + \|\hat{\phi}_2\| \right)^2 \|\omega_2(u_d,\alpha)\|^2 + \epsilon_{N2} \left\| \frac{\partial \epsilon_1(v,\alpha)}{\partial v} \right\|. \quad (7.30)
\end{aligned}
$$

Note that the bounds ϕ_A and ϕ_C of the parameter vectors ϕ_a^* and ϕ_c^* are not assumed to be known. With some algebraic simplifications, we get

$$|\epsilon(t)| \leq a_1\|\hat{\phi}_2\|^3 + a_2\|\hat{\phi}_2\|^2 + a_3\|\hat{\phi}_2\| + a_4\|\hat{\phi}_2\|\|\hat{\phi}_1\| + a_5\|\hat{\phi}_1\| + a_6 = \xi^T\Phi, \tag{7.31}$$

where $\xi = [a_1, \ldots, a_6]^T$ is a vector of bounded constants (not assumed to be known and will be updated by adaptive laws) defined using the bounds in assumption (A1), and $\Phi = \left[\|\hat{\phi}_2\|^3,\ \|\hat{\phi}_2\|^2,\ \|\hat{\phi}_2\|,\ \|\hat{\phi}_2\|\|\hat{\phi}_1\|,\ \|\hat{\phi}_1\|,\ 1\right]^T$ is a known function vector. The vector ξ is bounded because $\|\omega_2(u_d, \alpha)\|$ and $\left\|\frac{\partial\omega_1(v,\alpha)}{\partial v}\right\|$ are bounded for bounded α. The approximation error is reduced because of the consideration of the second ordered term. We use an estimator $\xi^T\Phi$ as

$$\hat{\epsilon} = \hat{\xi}^T\Phi, \tag{7.32}$$

where $\hat{\xi} = [a_1, \ldots, a_6]^T$ is adaptively updated.

The modification of one nonlinearity function depends on the other, and therefore the adaptive update laws to be developed for $\hat{\phi}_1^T$ and $\hat{\phi}_2^T$ should be mutually coupled, such that the parameter errors $\tilde{\phi}_1^T$ and $\tilde{\phi}_2^T$ are bounded and closed-loop stability is ensured. An explicit expression for ξ is not needed to be known and is updated through an adaptive law.

7.3 Nonlinear Aircraft Control Law Development

Next, we utilize the nonlinear aircraft dynamics model described in Chap. 3. The BANTAM configuration for operation under wide range of AoA (both low and high AoA) consists of arrays of SJAs (six shaded regions), and a leading edge jet array is utilized.

The jets at the trailing edge are activated while operating at low AoA, and the leading edge actuator array is operated while operating at high AoA. At intermediate AoA (for which detailed wind tunnel tests have not yet been conducted), one possible approach is to optimally select the jets from both the regions of the wing.

The states are

$$x = \left[p\ q\ r\ \gamma\ \chi\ \mu\ V\ \alpha\ \beta\right]^T, \tag{7.33}$$

where p, q, r are the roll, pitch and yaw rates, V is relative aircraft velocity, α is the angle of attack, β is the sideslip angle, γ, χ are the flight-path angles which are assumed to be positive when the aircraft is climbing and μ is the bank angle.

For the controller formulation, we make the following assumption:

(A2) the velocity V and the angle γ do not approach 0 and $\pm 90°$ respectively, and $-25° \leq \alpha \leq 25°$, so that the nonlinear aircraft dynamics avoid singularities.

The velocity control and error dynamics are already shown in Chap. 3. From the velocity dynamics in (3.1), for velocity control, and the desired thrust T as a control command, the velocity error dynamics is given by

$$\dot{\tilde{V}} = -K_V \tilde{V}. \tag{7.34}$$

With thrust T, the flight path angle dynamics (3.4), (3.5) are

$$\dot{\gamma} = f_\gamma + g_{\gamma,u}, \quad \dot{\chi} = f_\chi + g_{\chi,u}, \tag{7.35}$$

where the known functions (with cs and tn representing cosine and tangent functions respectively) are

$$f_\gamma = \frac{T\,\mathrm{cs}\,\alpha\,\mathrm{sn}\,\beta\,\mathrm{sn}\,\mu - mg\,\mathrm{cs}\,\gamma}{mV}, \quad f_\chi = -\frac{T\,\mathrm{cs}\,\alpha\,\mathrm{sn}\,\beta\,\mathrm{cs}\,\mu}{mV\,\mathrm{cs}\,\gamma},$$

and the functions $g_{\gamma,u}$ and $g_{\chi,u}$ are given by

$$g_{\gamma,u} = \frac{\mathrm{cs}\,\mu}{mV}u + \frac{T\,\mathrm{sn}\,\alpha\,\mathrm{cs}\,\mu}{mV}, \quad g_{\chi,u} = \frac{\mathrm{sn}\,\mu}{mV\,\mathrm{cs}\,\gamma}u + \frac{T\,\mathrm{sn}\,\alpha\,\mathrm{sn}\,\mu}{mV\,\mathrm{cs}\,\gamma}. \tag{7.36}$$

We generate μ_r, α_r from

$$\begin{aligned}
\frac{\mathrm{cs}\,\mu_r}{mV}\left(u_d + \hat{\xi}^T\Phi + T\,\mathrm{sn}\,\alpha_r\right) &= \dot{\gamma}_r - f_\gamma - K_\gamma\tilde{\gamma}, \; K_\gamma > 0,\\
\frac{\mathrm{sn}\,\mu_r}{mV\,\mathrm{cs}\,\gamma}\left(u_d + \hat{\xi}^T\Phi + T\,\mathrm{sn}\,\alpha_r\right) &= \dot{\chi}_r - f_\chi - K_\chi\tilde{\chi}, \; K_\chi > 0,
\end{aligned} \tag{7.37}$$

where χ_r and γ_r are the available flight path angles, $\tilde{\gamma} = \gamma - \gamma_r$, $\tilde{\chi} = \chi - \chi_r$, are flight path errors. The angles μ_r, α_r are used as control signals for the wind axis loop. From (7.37), μ_r and α_r are given by

$$\mu_r = \arctan\left(\frac{\mathrm{cs}\,\gamma\,(\dot{\chi}_r - f_\chi - K_\chi\tilde{\chi})}{\dot{\gamma}_r - f_\gamma - K_\gamma\tilde{\gamma}}\right), \alpha_r = \arcsin\left(\frac{mVF_\gamma - u_d - \hat{\xi}^T\Phi}{T}\right) \tag{7.38}$$

where $F_\gamma = \frac{\dot{\gamma}_r - f_\gamma - K_\gamma\tilde{\gamma}}{\mathrm{cs}\,\mu_r}$. The validity of the solution of α_r is determined as per

$$\left|mVF_\gamma - u_d - \hat{\xi}^T\Phi\right| < \kappa\,|T|, \tag{7.39}$$

where κ is a constant determined by the allowable values of $\alpha_r \in (-25°, 25°)$, $\kappa = 0.1736$.

Using (7.35) and (7.37), the dynamics of $\tilde{\gamma}, \tilde{\chi}$ are

$$\dot{\tilde{\gamma}} = -K_\gamma \tilde{\gamma} + \frac{T\,(\mathrm{sn}\,\alpha\,\mathrm{cs}\,\mu - \mathrm{sn}\,\alpha_r\,\mathrm{cs}\,\mu_r)}{mV} + \frac{\mathrm{cs}\,\mu}{mV}u - \frac{\mathrm{cs}\,\mu_r}{mV}\left(u_d + \hat{\xi}^T\Phi\right),$$
$$\dot{\tilde{\chi}} = -K_\chi \tilde{\chi} + \frac{T\,(\mathrm{sn}\,\alpha\,\mathrm{sn}\,\mu - \mathrm{sn}\,\alpha_r\,\mathrm{sn}\,\mu_r)}{mV\,\mathrm{cs}\,\gamma} + \frac{\mathrm{sn}\,\mu}{mV\,\mathrm{cs}\,\gamma}u - \frac{\mathrm{sn}\,\mu_r}{mV\,\mathrm{cs}\,\gamma}\left(u_d + \hat{\xi}^T\Phi\right). \tag{7.40}$$

Using Taylor series expansions, we have

$$\begin{aligned} \mathrm{cs}\,\mu &= \mathrm{cs}\,\mu_r + O_1(\tilde{\mu}, \mu_r), \quad \mathrm{sn}\,\mu = \mathrm{sn}\,\mu_r + O_2(\tilde{\mu}, \mu_r), \\ \mathrm{sn}\,\alpha\,\mathrm{cs}\,\mu &= \mathrm{sn}\,\alpha_r\,\mathrm{cs}\,\mu_r + \tilde{\alpha}\,\mathrm{cs}\,\alpha_r\,\mathrm{cs}\,\mu_r - \tilde{\mu}\,\mathrm{sn}\,\alpha_r\,\mathrm{sn}\,\mu_r + O_3(\tilde{\mu}, \tilde{\alpha}), \\ \mathrm{sn}\,\alpha\,\mathrm{sn}\,\mu &= \mathrm{sn}\,\alpha_r\,\mathrm{sn}\,\mu_r + \tilde{\mu}\,\mathrm{sn}\,\alpha_r\,\mathrm{cs}\,\mu_r + \tilde{\alpha}\,\mathrm{sn}\,\mu_r\,\mathrm{cs}\,\alpha_r + O_4(\tilde{\mu}, \tilde{\alpha}), \end{aligned} \tag{7.41}$$

where $\tilde{\mu} = \mu - \mu_r, \tilde{\alpha} = \alpha - \alpha_r$, and $O_i(\tilde{\mu}, \mu_r), i = 1, 2,\ O_i(\tilde{\mu}, \tilde{\alpha}), i = 3, 4$ contain second and higher order arguments of $\tilde{\mu}$, such that

$$\lim_{|\tilde{\mu}|\to 0} \frac{O_i(\tilde{\mu}, \mu_r)}{|\tilde{\mu}|} = 0, i = 1, 2, \qquad \lim_{|\tilde{\mu}|,|\tilde{\alpha}|\to 0} \frac{O_i(\tilde{\mu}, \tilde{\alpha})}{\sqrt{|\tilde{\mu}|^2 + |\tilde{\alpha}|^2}} = 0, i = 3, 4.$$

Using (7.40) and (7.41), the approximated flight path angle errors are given by

$$\begin{aligned} \dot{\tilde{\gamma}} &= -K_\gamma \tilde{\gamma} + \frac{\mathrm{cs}\,\mu}{mV}u - \frac{\mathrm{cs}\,\mu_r}{mV}\left(u_d + \hat{\xi}^T\Phi\right) + \frac{T\,(\tilde{\alpha}\,\mathrm{cs}\,\alpha_r\,\mathrm{cs}\,\mu_r - \tilde{\mu}\,\mathrm{sn}\,\mu_r\,\mathrm{sn}\,\alpha_r)}{mV} \\ &\quad + \frac{O_1(\tilde{\mu}, \mu_r)u_d + O_3(\tilde{\mu}, \tilde{\alpha})T}{mV}, \\ \dot{\tilde{\chi}} &= -K_\chi \tilde{\chi} + \frac{\mathrm{sn}\,\mu}{mV\,\mathrm{cs}\,\gamma}u - \frac{\mathrm{sn}\,\mu_r}{mV\,\mathrm{cs}\,\gamma}\left(u_d + \hat{\xi}^T\Phi\right) \\ &\quad + \frac{T\,(\tilde{\alpha}\,\mathrm{sn}\,\mu_r\,\mathrm{cs}\,\alpha_r + \tilde{\mu}\,\mathrm{cs}\,\mu_r\,\mathrm{sn}\,\alpha_r)}{mV\,\mathrm{cs}\,\gamma} + \frac{O_2(\tilde{\mu}, \mu_r)u_d + O_4(\tilde{\mu}, \tilde{\alpha})T}{mV\,\mathrm{cs}\,\gamma}. \end{aligned} \tag{7.42}$$

With thrust T, the wind axis angle dynamics (3.2)–(3.3), are given by

$$\dot{\mu} = f_\mu + g_{\mu,u}, \quad \dot{\alpha} = f_\alpha + g_{\alpha,u}, \quad \dot{\beta} = f_\beta + g_{\beta,u}, \tag{7.43}$$

where the known functions f_μ, f_α and f_β are

$$\begin{aligned} f_\mu &= \frac{T}{mV}\,(\mathrm{sn}\,\alpha\,\mathrm{tn}\,\gamma\,\mathrm{sn}\,\mu + \mathrm{sn}\,\alpha\,\mathrm{tn}\,\beta - \mathrm{cs}\,\alpha\,\mathrm{tn}\,\gamma\,\mathrm{cs}\,\mu\,\mathrm{sn}\,\beta) - \frac{g}{V}\,\mathrm{cs}\,\alpha\,\mathrm{cs}\,\mu\,\mathrm{tn}\,\beta, \\ f_\alpha &= \frac{1}{mV\,\mathrm{cs}\,\beta}\,(-T\,\mathrm{sn}\,\alpha + mg\,\mathrm{cs}\,\gamma\,\mathrm{cs}\,\mu) - (p\,\mathrm{cs}\,\alpha + r\,\mathrm{sn}\,\alpha)\,\mathrm{tn}\,\beta, \\ f_\beta &= \frac{1}{mV}\,(-T\,\mathrm{sn}\,\beta\,\mathrm{cs}\,\alpha + mg\,\mathrm{cs}\,\gamma\,\mathrm{sn}\,\mu), \end{aligned} \tag{7.44}$$

and the functions $g_{\mu,u}$, $g_{\alpha,u}$ and $g_{\beta,u}$ are given by

$$g_{\mu,u} = \frac{p\,\mathrm{cs}\,\alpha + r\,\mathrm{sn}\,\alpha}{\mathrm{cs}\,\beta} + \frac{\mathrm{tn}\,\beta + \mathrm{tn}\,\gamma\,\mathrm{sn}\,\mu}{mV}u,$$

$$g_{\alpha,u} = -\frac{1}{mV\,\mathrm{cs}\,\beta}u + q, \quad g_{\beta,u} = p\,\mathrm{sn}\,\alpha - r\,\mathrm{cs}\,\alpha. \tag{7.45}$$

For wind axis angles control, we determine the signals p_r, q_r, r_r, from

$$\frac{p_r\,\mathrm{cs}\,\alpha + r_r\,\mathrm{sn}\,\alpha}{\mathrm{cs}\,\beta} + \frac{\mathrm{tn}\,\beta + \mathrm{tn}\,\gamma\,\mathrm{sn}\,\mu}{mV}\left(u_d + \hat{\xi}^T\Phi\right) = -f_\mu + \dot{\mu}_r - K_\mu\tilde{\mu}$$
$$+\frac{T\,\mathrm{sn}\,\alpha_r}{mV}\left(-\tilde{\chi}\frac{\mathrm{cs}\,\mu_r}{\mathrm{cs}\,\gamma} + \tilde{\gamma}\,\mathrm{sn}\,\mu_r\right),$$

$$q_r - \frac{1}{mV\,\mathrm{cs}\,\beta}u_d = -f_\alpha + \dot{\alpha}_r - K_\alpha\tilde{\alpha} + \frac{1}{mV\,\mathrm{cs}\,\beta}\hat{\xi}^T\Phi - \frac{T\,\mathrm{cs}\,\alpha_r}{mV}\left(\tilde{\chi}\frac{\mathrm{sn}\,\mu_r}{\mathrm{cs}\,\gamma} + \tilde{\gamma}\,\mathrm{cs}\,\mu_r\right),$$
$$p_r\,\mathrm{sn}\,\alpha - r_r\,\mathrm{cs}\,\alpha = -f_\beta + \dot{\beta}_r - K_\beta\tilde{\beta}, \tag{7.46}$$

where μ_r, α_r are generated from (7.38), and are therefore available now for wind axis control, and $K_\mu, K_\alpha, K_\beta > 0$ are constant gains, $\tilde{\mu} = \mu - \mu_r$, $\tilde{\alpha} = \alpha - \alpha_r$, $\tilde{\beta} = \beta - \beta_r$ are the wind axis angle errors. The particular form of (7.46) enables the error dynamics.

Using (7.43)–(7.46), the dynamic equations for wind axis errors $\tilde{\mu} = \mu - \mu_r$, $\tilde{\alpha} = \alpha - \alpha_r$ and $\tilde{\beta} = \beta - \beta_r$, are given by

$$\dot{\tilde{\mu}} = -K_\mu\tilde{\mu} - \frac{T\,\mathrm{sn}\,\alpha_r}{mV}\left(\tilde{\chi}\frac{\mathrm{cs}\,\mu_r}{\mathrm{cs}\,\gamma} - \tilde{\gamma}\,\mathrm{sn}\,\mu_r\right) + \frac{\tilde{p}\,\mathrm{cs}\,\alpha + \tilde{r}\,\mathrm{sn}\,\alpha}{\mathrm{cs}\,\beta}$$
$$+\frac{\mathrm{tn}\,\beta + \mathrm{tn}\,\gamma\,\mathrm{sn}\,\mu}{mV}\left(u - u_d - \hat{\xi}^T\Phi\right),$$
$$\dot{\tilde{\alpha}} = -K_\alpha\tilde{\alpha} + \tilde{q} + \frac{T\,\mathrm{cs}\,\alpha_r}{mV}\left(-\tilde{\chi}\frac{\mathrm{sn}\,\mu_r}{\mathrm{cs}\,\gamma} - \tilde{\gamma}\,\mathrm{cs}\,\mu_r\right) - \frac{1}{mV\,\mathrm{cs}\,\beta}\left(u - u_d - \hat{\xi}^T\Phi\right),$$
$$\dot{\tilde{\beta}} = -K_\beta\tilde{\beta} + \tilde{p}\,\mathrm{sn}\,\alpha - \tilde{r}\,\mathrm{cs}\,\alpha, \tag{7.47}$$

where $\tilde{p} = p - p_r$, $\tilde{q} = q - q_r$, and $\tilde{r} = r - r_r$ are the body axis rate errors.

The body axis equations given by (3.6)–(3.8), are

$$\dot{p} = f_p + g_{p,u}, \quad \dot{q} = f_q + g_{q,u}, \quad \dot{r} = f_r + g_{r,u}, \tag{7.48}$$

where the known functions are given by

$$f_p = (c_1 r + c_2 p)q + c_3\bar{l}_a(t),$$
$$f_q = c_4 pr + c_5\bar{m}_a(t),$$
$$f_r = (c_6 p - c_2 r)q + c_7\bar{l}_a(t), \tag{7.49}$$

for known aerodynamic moments $\bar{l}_a(t), \bar{m}_a(t), \bar{n}_a(t)$, the functions $g_{p,u}, g_{q,u}, g_{r,u}$, are

$$g_{p,u} = c_3 u_1, \ g_{q,u} = c_5 u_2, \ g_{r,u} = c_7 u_3,$$

where the stability axis roll, pitch and yaw moments are $u_1 = a_1 u = \bar{l}(t) - \bar{l}_a(t)$, $u_2 = a_2 u = \bar{m}(t) - \bar{m}_a(t)$, $u_3 = a_3 u = \bar{n}(t) - \bar{n}_a(t)$ respectively, and a_1, a_2, a_3 are the parameters determining allocation of control authority.

The desired lift forces $u_d(t)$, are generated from the equations

$$\begin{aligned} c_3\left(u_{d1} + a_1\hat{\xi}^T\Phi\right) &= -f_p - K_p\tilde{p} + \dot{p}_r - \left(\tilde{\mu}\frac{\operatorname{cs}\alpha}{\operatorname{cs}\beta} + \tilde{\beta}\operatorname{sn}\alpha\right), \\ c_5\left(u_{d2} + a_2\hat{\xi}^T\Phi\right) &= -f_q + \dot{q}_r - K_q\tilde{q} - \tilde{\alpha}, \\ c_7\left(u_{d3} + a_3\hat{\xi}^T\Phi\right) &= -f_r - K_r\tilde{r} + \dot{r}_r - \left(\tilde{\mu}\frac{\operatorname{sn}\alpha}{\operatorname{cs}\beta} - \tilde{\beta}\operatorname{cs}\alpha\right), \end{aligned} \tag{7.50}$$

where p_r, q_r, r_r are generated from (7.46), and are therefore available now for body axis rate control, $K_p, K_q, K_r > 0$, are constant gains, and $u_d = \frac{1}{a_1}u_{d1} = \frac{1}{a_2}u_{d2} = \frac{1}{a_3}u_{d3}$. In (7.50), additional tracking error terms containing $\tilde{\mu}, \tilde{\alpha}, \tilde{\beta}$ are inserted to ensure Lyapunov stability.

Using (7.48)–(7.50), the equations for the body axis rate errors $\tilde{p} = p - p_r, \tilde{q} = q - q_r, \tilde{r} = r - r_r$, are

$$\begin{aligned} \dot{\tilde{p}} &= -\left(\tilde{\mu}\frac{\operatorname{cs}\alpha}{\operatorname{cs}\beta} + \tilde{\beta}\operatorname{sn}\alpha\right) + c_3 a_1\left(u - u_d - \hat{\xi}^T\Phi\right) - K_p\tilde{p}, \\ \dot{\tilde{q}} &= c_5 a_2\left(u - u_d - \hat{\xi}^T\Phi\right) - K_q\tilde{q} - \tilde{\alpha}, \\ \dot{\tilde{r}} &= \tilde{\beta}\operatorname{cs}\alpha - \tilde{\mu}\frac{\operatorname{sn}\alpha}{\operatorname{cs}\beta} - K_r\tilde{r} + c_7 a_3\left(u - u_d - \hat{\xi}^T\Phi\right). \end{aligned} \tag{7.51}$$

7.4 Adaptation Scheme

In this section, we formulate adaptive laws to update the parameter estimates $\hat{\phi}_1(t), \hat{\phi}_r(t)$, and $\hat{\xi}(t)$ so that the control objective is achievable. Parameter projection is used to guarantee the boundedness of these parameters by lower bounds. We assume that the following bounds of the components of the true nonlinearity parameters ϕ_a^*, ϕ_r^*, and ξ are known:

$$\phi_{1a}^b \le \phi_{1a}^*, \dots, \phi_{k_1 a}^b \le \phi_{k_1 a}^*, \ \phi_{1c}^b \le \phi_{1c}^*, \dots, \phi_{k_2 a}^b \le \phi_{k_2 a}^*,$$

$$\xi_1^b \le \xi_1 \le \xi_1^d, \dots, \xi_4^b \le \xi_4 \le \xi_1^d,$$

for some known constants $\phi^b_{1a}, \ldots, \phi^b_{k_1 a} > 0,\ k_1 = (R_1+1)(M_1+1),\ \phi^b_{1c}, \ldots, \phi^b_{k_2 c} > 0,\ k_2 = (R_2+1)(M_2+1),\ \xi^b_1, \ldots, \xi^b_4$, and $\xi^d_1, \ldots, \xi^d_4$.

Initial estimates $\hat{\phi}_1(0)$, $\hat{\phi}_2(0)$ and $\hat{\xi}(0)$ of ϕ^*_a, ϕ^*_c, and ξ respectively are chosen such that

$$\phi^b_{1a} < \hat{\phi}_{1a}, \ldots, \phi^b_{k_1 a} \le \hat{\phi}_{k_1 a},\ \ \phi^b_{1c} < \hat{\phi}_{1c}, \ldots, \phi^b_{k_2 c} \le \hat{\phi}_{k_2 c},$$
$$\xi^b_1 \le \hat{\xi}_1 \le \xi^d_1, \ldots, \xi^b_4 \le \hat{\xi}_4 \le \xi^d_1. \tag{7.52}$$

Adaptive laws: We choose the adaptive laws as

$$\begin{aligned}\dot{\hat{\phi}}_1(t) &= g_1(t) + h_1(t),\\ \dot{\hat{\phi}}_2(t) &= g_2(t) + h_2(t),\\ \dot{\hat{\xi}}(t) &= \Gamma_3 \Phi |e(t)| + h_\xi(t),\end{aligned} \tag{7.53}$$

where $g_1(t)$ and $g_2(t)$ are adaptation functions given by

$$g_1(t) = \Gamma_1 e(t) \left(\frac{\partial \omega_1(v,\alpha)}{\partial v} - \frac{1}{2}\frac{\partial^2 \omega_1(v,\alpha)}{\partial v^2} \hat{\phi}^T_2 \omega_2(u_d,\alpha) \right) \hat{\phi}^T_2 \omega_2(u_d,\alpha),$$

$$g_2(t) = -\Gamma_2 e(t) \left(\frac{\partial \omega_1(v,\alpha)}{\partial v} - \frac{1}{2}\frac{\partial^2 \omega_1(v,\alpha)}{\partial v^2} \hat{\phi}^T_2 \omega_2(u_d,\alpha) \right) \hat{\phi}^T_2 \omega_2(u_d,\alpha),$$

and Γ_i, $i = 1, 2, 3$, are adaptation gain matrices such that $\Gamma_i = \Gamma_i^T > 0$, and

$$\begin{aligned} e(t) &= -\frac{\tilde{\alpha}}{mV \operatorname{cs}\beta} + \frac{\operatorname{tn}\beta + \operatorname{tn}\gamma \operatorname{sn}\mu}{mV}\tilde{\mu} + \frac{\operatorname{sn}\mu}{mV \operatorname{cs}\gamma}\tilde{\chi}\\ &\quad + \frac{\operatorname{cs}\mu}{mV}\tilde{\gamma} + c_3 a_1 \tilde{p} + c_5 a_2 \tilde{q} + c_5 a_3 \tilde{r}, \end{aligned} \tag{7.54}$$

and $h_1(t)$, $h_2(t)$ and $h_\xi(t)$ are the projection functions which ensure that the estimates of the parameters are constrained in a desired region, given as

$$h_{ij}(t) = \begin{cases} 0 & \text{if } \ \hat{\phi}_{ij}(t) > \phi^b_{ij}, \ \text{ or} \\ & \text{if } \ \hat{\phi}_{ij}(t) = \phi^b_{ij} \text{ and } g_{ij}(t) \ge 0, \\ -g_{ij}(t) & \text{otherwise,} \end{cases} \tag{7.55}$$

where $i = 1, \ldots, (R_1+1)(M_1+1)$ for $j = a$, and $i = 1, \ldots, (R_2+1)(M_2+1)$ for $j = c$, and

$$h_{\xi k}(t) = \begin{cases} 0 & \text{if } \hat{\xi}_k(t) \in \left(\xi_k^b, \xi_k^d\right), \text{ or} \\ & \text{if } \hat{\xi}_k(t) = \xi_k^b, \text{ and } \Gamma_3 \Phi e(t) \le 0, \text{ or} \\ & \text{if } \hat{\xi}_k(t) = \xi_k^d, \text{ and } \Gamma_3 \Phi e(t) \ge 0, \\ \Gamma_3 \Phi e(t) & \text{otherwise,} \end{cases} \tag{7.56}$$

where $k = 1, \ldots, 4,\ h_\xi = [h_{\xi 1}, \ldots, h_{\xi 4}]^T$.

7.5 Performance Analysis

In this section, we analyze the aircraft dynamic system stability performance.

Theorem 7.1 *Under assumptions (A1) and (A2), the error dynamics (7.34), (7.42), (7.47), and (7.51) from state feedback control law governed by (7.38), (7.46) and (7.50), with parameters updated from the adaptive laws (7.53), and an error function given by (7.54) guarantee that the overall closed-loop system is semi-globally stable in the sense that all of the signals in the closed-loop system are bounded and $\tilde{x}$ eventually converges to the compact set $S(\delta^2)$ given by*

$$S(\delta^2) = \left\{ \tilde{x} \,\middle|\, \int_t^{t+T} \|\tilde{x}(\tau)\|_2^2 \le \delta^2 T + c_0 \right\}, \tag{7.57}$$

that is, $\tilde{x}$ is δ^2-small in the mean squared sense, and c_0 is a positive constant, for $\tilde{x} = \left[V\ \tilde{\gamma}\ \tilde{\chi}\ \tilde{\mu}\ \tilde{\alpha}\ \tilde{\xi}\ \tilde{p}\ \tilde{q}\ \tilde{r}\right]^T$.

Proof Consider the positive definite function

$$W = \frac{1}{2}\tilde{x}^T\tilde{x} + \frac{1}{2}\tilde{\phi}_1^T\Gamma_1^{-1}\tilde{\phi}_1 + \frac{1}{2}\tilde{\phi}_2^T\Gamma_2^{-1}\tilde{\phi}_2 + \frac{1}{2}\tilde{\xi}^T\Gamma_3^{-1}\tilde{\xi}, \tag{7.58}$$

where $\tilde{\xi}(t) = \xi - \hat{\xi}(t)$. Using (7.34), (7.42), (7.47), and (7.51), and differentiating W, we have

$$\dot{W} = -K_x\tilde{x}^2 + e(t)\left(u - u_d - \hat{\xi}^T\Phi\right) + \Delta - \tilde{\phi}_1^T\Gamma_1^{-1}\dot{\hat{\phi}}_1 - \tilde{\phi}_2^T\Gamma_2^{-1}\dot{\hat{\phi}}_2 - \tilde{\xi}^T\Gamma_3^{-1}\dot{\hat{\xi}}, \tag{7.59}$$

where $K_x = \left[K_V\ K_\chi\ K_\gamma\ K_\mu\ K_\alpha\ K_\beta\ K_p\ K_q\ K_r\right]$, Δ represents the third order (and above) error terms given as

$$\Delta = \frac{\tilde{\gamma}}{mV}\left(O_1(\tilde{\mu})u_d + O_3(\tilde{\mu}, \tilde{\alpha})T\right) + \frac{\tilde{\chi}}{mV\,\mathrm{cs}\,\gamma}\left(O_2(\tilde{\mu})u_d + O_4(\tilde{\mu}, \tilde{\alpha})T\right), \tag{7.60}$$

where $\tilde{\mu}$, $\tilde{\alpha}$ are flight path angle errors, and $O_i(\tilde{\mu})$, $i = 1, 2$, $O_i(\tilde{\mu}, \tilde{\alpha})$, $i = 3, 4$, represent higher order error terms such that

$$\Delta \leq \eta_1 \tilde{\gamma} + \eta_2 \tilde{\chi}, \tag{7.61}$$

for computable bounds η_1 and η_2. Substituting the control error (7.28) and the adaptive laws (7.53), we have

$$\begin{aligned}
\dot{W} \leq & -K_x \tilde{x}^2 - \tilde{\phi}_1^T \Gamma_1^{-1} \dot{\hat{\phi}}_1 - \tilde{\phi}_2^T \Gamma_2^{-1} \dot{\hat{\phi}}_2 + \delta_1 \tilde{\gamma} + \delta_2 \tilde{\chi} + e(t)\epsilon(t) \\
& + e(t) \left(\frac{\partial \omega_1(v,\alpha)}{\partial v} - \frac{1}{2} \frac{\partial^2 \omega_1(v,\alpha)}{\partial v^2} \hat{\phi}_2^T \omega_2(u_d,\alpha) \right) \hat{\phi}_2^T \omega_2(u_d,\alpha) \\
& - e(t) \left(\frac{\partial \omega_1(v,\alpha)}{\partial v} - \frac{1}{2} \frac{\partial^2 \omega_1(v,\alpha)}{\partial v^2} \hat{\phi}_2^T \omega_2(u_d,\alpha) \right) \hat{\phi}_1^T \omega_2(u_d,\alpha) - e(t)\hat{\xi}^T \Phi \\
& - \tilde{\xi}^T \Phi |e(t)| - \tilde{\xi}^T \Gamma_3^{-1} h_\xi(t) \\
\leq & -K_x \tilde{x}^2 + e(t)\epsilon(t) - \tilde{\phi}_1^T \Gamma_1^{-1} h_1(t) - \tilde{\phi}_2^T \Gamma_2^{-1} h_2(t) - \hat{\xi}^T \Phi e(t) - \tilde{\xi}^T \Phi |e(t)| \\
& - \tilde{\xi}^T \Gamma_3^{-1} h_\xi(t) + \eta_1 \tilde{\gamma} + \eta_2 \tilde{\chi}.
\end{aligned} \tag{7.62}$$

With appropriate and known initial conditions and from the parameter projection functions $h_1(t)$ and $h_2(t)$ which maintain the parameters in the desired bounds, it follows that (7.52) is satisfied, and

$$\tilde{\phi}_a^T(t) h_1(t) \geq 0, \quad \tilde{\phi}_c^T(t) h_2(t) \geq 0, \quad \tilde{\xi}^T h_\xi(t) \geq 0.$$

With this property and since $|\epsilon(t)| \leq \xi^T \Phi$, (7.62) reduces to

$$\dot{W} \leq -K_x \tilde{x}^2 + |e(t)| \xi^T \Phi - \hat{\xi}^T \Phi |e(t)| - \tilde{\xi}^T \Phi |e(t)| + \eta_1 \tilde{\gamma} + \eta_2 \tilde{\chi}. \tag{7.63}$$

Using the inequality

$$ab \leq \zeta^2 x^2 + \frac{1}{4\zeta^2} y^2, \ \forall \zeta \neq 0,$$

for $\zeta^2 = \frac{K_\gamma}{2}, \frac{K_\chi}{2}$ respectively, and $|\eta_i| \leq \sigma_i, i = 1, 2$, we have

$$\begin{aligned}
-K_\gamma \tilde{\gamma}^2 + \eta_1 \tilde{\gamma} &\leq -\frac{K_\gamma}{2} \tilde{\gamma}^2 + \frac{1}{2K_\gamma} \sigma_1^2, \\
-K_\chi \tilde{\chi}^2 + \eta_2 \tilde{\chi} &\leq -\frac{K_\chi}{2} \tilde{\chi}^2 + \frac{1}{2K_\chi} \sigma_2^2.
\end{aligned} \tag{7.64}$$

Substituting (7.64) in (7.63), we find that

$$\dot{W} \leq -K \tilde{x}^2 + \frac{1}{2 \min\{K_\gamma, K_\chi\}} \|\sigma\|^2, \tag{7.65}$$

where $\sigma = [\sigma_1, \sigma_2]^T$ and $K = \left[K_V \ \frac{K_\chi}{2} \ \frac{K_\gamma}{2} \ K_\mu \ K_\alpha \ K_\beta \ K_p \ K_q \ K_r\right]$. Integrating both sides over the time interval $[t, t+T]$, we have

$$\int_t^{t+T} \dot{W}(\tau)d\tau \leq -\|K\| \int_t^{t+T} \|\tilde{x}(\tau)\|_2^2 d\tau + \frac{\|\sigma\|^2}{2\min\{K_\gamma, K_\chi\}} T. \tag{7.66}$$

Therefore, we have

$$\int_t^{t+T} \|\tilde{x}(\tau)\|_2^2 d\tau \leq \delta^2 T + c_0, \forall t \geq 0, \text{ and any } T \geq 0, \tag{7.67}$$

where c_0 and δ^2 are given as

$$c_0 = \sup_{t \geq 0} \frac{1}{\|K\|} \left(W(t) - W(t+T)\right), \ \delta^2 = \frac{\sigma^2}{2\|K\| \min\{K_\gamma, K_\chi\}}. \tag{7.68}$$

Therefore, we have $\tilde{x}(\tau) \in S(\delta^2)$, where δ can be adjusted by appropriately choosing the design parameters and where $\tilde{\phi}_1(t), \tilde{\phi}_2(t)$ are also uniformly bounded. We have presented an adaptive inverse approximation control scheme, which operates within a coverage region with the objective of causing the tracking error terms of the aircraft to converge to a small neighborhood. The size of this neighborhood is determined (in part) by the residual approximation error. $\bigtriangledown$

This chapter presented a new adaptive control technology for control of synthetic jet actuators for aircraft applications in which system and actuation uncertainties are a major concern. An adaptive compensation scheme is presented for SJA nonlinearities at a wide range of AoA. The dependence of the structure of the nonlinearity model on AoA is not precisely known. A new technique for inversion of an unknown invertible function is presented.

7.6 Concluding Remarks

The SJ actuation technologies aim to provide a new perspective to aerodynamic control systems applications. There are, however, associated control challenges because the SJAs have inherent nonlinearities. The AIC technique is cost-effective because it is augmentable to the existing nominal flight controllers without requiring wholesale redeployment of control algorithms simply because of the usage of these new SJAs. This work focuses on the development and analysis of an adaptive feedback control system for flight control with SJAs deployed.

In Chap. 2, the physical characteristics of a SJA are described. The mathematical model of the actuator is also presented. An AIC cancels the effects of the uncertain actuator nonlinearities. The general control objective to ensure desired tracking performance and closed-loop stability in the presence of unknown system parameters is developed in subsequent chapters at different conditions of AoA and different aircraft dynamic models. Based on this control objective, adaptive compensation of SJAs is developed for several typical classes of nonlinear systems. Adaptive compensation of SJA nonlinearities is studied for a linear aircraft model.

The AIC technique, as has been known in the last few decades, is applicable when the actuator uncertainties are not known but the structure of the actuator function is known. However, in the present work, we encounter a situation wherein the SJA function is signal-dependent (or state-dependent) as the actuation function itself changes with change in the angle of attack of the aircraft. The functional form of the SJA model is unavailable (or imprecisely known) due to this feature of signal-dependency. Using twin-functional approximators, we have developed a new adaptation technique for this specific situation. We have successfully applied this method with two neural networks on a 3-DoF aircraft system (Chap. 5), two Taylor series approximators on a general class of nonlinear system (Chap. 6), and also deployed it on a 6-DoF aircraft model along with SJAs (Chap. 7).

A unified adaptive compensation scheme is developed for controlling state-dependent actuator nonlinearities in a class of nonlinear system. Approximation of a nonlinearly parametrized actuator model by a linearly parametrized function is performed. Adaptive inversion of actuator nonlinearities for the linearly approximated model is performed by adopting another approximator in the feedforward path. A nonlinear state feedback control law is designed for controlling the nonlinear dynamics. An adaptive control technology for control of state-dependent SJA nonlinearities in an aircraft flight control system, is developed.

Adaptive compensation is a promising approach to the problem of signal dependent actuator nonlinearities. The study in this dissertation shows some encouraging results in adaptive actuator compensation for linear and nonlinear systems. At the end of this dissertation, several research topics including both theoretical study and potential applications are listed here as some future directions in this area.

References

1. Deb, D., Tao, G., Burkholder, J., Smith, D.: An adaptive inverse control scheme for a synthetic jet actuator model. In: Proceedings of the 2005, American Control Conference (2005). https://doi.org/10.1109/acc.2005.1470367
2. Deb, D., Tao, G., Burkholder, J., Smith, D.: Adaptive synthetic jet actuator compensation for a nonlinear tailless aircraft model at low angles of attack. In: 2006 American Control Conference (2006). https://doi.org/10.1109/acc.2006.1655376
3. Deb, D., Tao, G., Burkholder, J., Smith, D.: An adaptive inverse control scheme for synthetic jet actuator arrays. In: Infotech@Aerospace (2005). https://doi.org/10.2514/6.2005-7170
4. Deb, D., Tao, G., Burkholder, J.: An adaptive inverse compensation scheme for signal-dependent actuator nonlinearities. In: 2007 46th IEEE Conference on Decision and Control (2007). https://doi.org/10.1109/cdc.2007.4434743

Correction to: Adaptive Compensation of Nonlinear Actuators for Flight Control Applications

Correction to:
D. Deb et al., *Adaptive Compensation of Nonlinear Actuators for Flight Control Applications*, Studies in Systems, Decision and Control 386, https://doi.org/10.1007/978-981-16-4161-9

The original version of the book was inadvertently published with an incorrect affiliation of the editor Dipankar Deb which has now been corrected as Institute of Infrastructure, Technology, Research and Management. The book has been updated with the change.

The updated version of the book can be found at
https://doi.org/10.1007/978-981-16-4161-9

D. Deb et al., *Adaptive Compensation of Nonlinear Actuators for Flight Control Applications*, Studies in Systems, Decision and Control 386,
https://doi.org/10.1007/978-981-16-4161-9_8